POCKET GUIDE
TO STARS &
PLANETS

POCKET GUIDE TO STARS & PLANETS

Ian Morison & Margaret Penston

First published in 2005 by New Holland Publishers

London • Cape Town • Sydney • Auckland

www.newhollandpublishers.com

86 Edgware Road, London W2 2EA, United Kingdom • 80 McKenzie Street, Cape Town
8001, South Africa • 14 Aquatic Drive, Frenchs Forest, NSW 2086, Australia • 218 Lake
Road, Northcote, Auckland, New Zealand

ISBN 1 84537 025 2

Although the publishers have made every effort to ensure that the information contained in this book was
meticulously researched and correct at the time of going to press, they
accept no responsibility for any inaccuracies, loss, injury or inconvenience sustained by any person using
this book as reference.

Publishing managers: Claudia dos Santos, Simon Pooley
• Commissioning editor: Alfred LeMaitre • Concept design: Geraldine Cupido
• Illustrations and design: Steven Felmore • Editor: Leizel Brown
• Production: Myrna Collins
Printed and bound in Singapore by Kyodo Printing Co (Singapore) (Pte) Ltd

2 4 6 8 10 9 7 5 3

CAPTIONS Half title: The Leonid meteor shower.
Full title: Sutherland Observatory, Northern Cape, South Africa – home to SALT (Southern
African Large Telescope). • This page: An artist's depiction of the Voyager satellite
exploring deep space, with the Milky Way as a backdrop. • Contents: Phases of a lunar eclipse

CONTENTS

1

THE APPEAL OF ASTRONOMY

This book will encourage and help you to make observations of the heavens. It will tell you what you should expect to see when observing the objects within our own solar system – the Moon and planets, meteors and comets. It will show you how to find your way around the constellations (for the Northern and Southern hemispheres) and describe how to find and observe 50 of the best objects in the sky as first introduced in the 'Astronomical A-List'. It will give you advice on a suitable tele-scope and how best to use it. Finally, it will give you sufficient background about the observable objects to enable you to understand what makes up the vast universe in which we live.

Opposite *An artist's representa-tion of the solar system seen against a backdrop of the stars in our galaxy. Here they are closer than in real life, with the beautiful ringed planet Saturn in the foreground.*

Astronomy as a hobby

Astronomy is a wonderful hobby with many facets. Some take pleasure in observing the Moon and planets, possible in even the most light-polluted cities; while others trek into the wilderness to find the darkest skies and track down faint galaxies whose light has taken many millions of years to reach us. Many join their local astronomy societies and enjoy the camaraderie of star parties and group observations of significant events such as meteor showers, bright comets and, more recently, the closest approach of Mars for 60,000 years as well as the Transit of Venus.

It is a great time to enter the hobby, and it need not be expensive. There is a wealth of high-quality telescopes available at surprisingly reasonable prices. In addition to this, superb magazines and Internet-based material are available to help you make the most of your observing time.

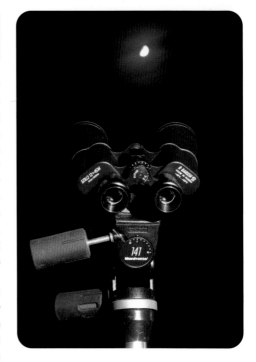

Above *A pair of 25x100 binoculars mounted on a tripod for absolute steadiness can serve as an excellent optical aid for celestial viewing.*

Primary equipment

There are only two basic pieces of equipment needed: a pair of binoculars and a telescope. A good pair of binoculars need not be expensive and really ought to be obtained first. Telescopes are available at a variety of prices and a newcomer might well expect that a large sum spent means that you'll see more. This is by no means the case! When, in 2003, Mars was at its closest to Earth in recorded history, every amateur society or observatory had a star party to witness it. At the party organized by the Jodrell Bank Observatory in the UK, there were over 30 telescopes and the best view of Mars was observed through the second cheapest one – an 8in (200mm) Newtonian on a simple mount.

A FINAL POINT

If you can join a local astronomy society, try to wait a while before buying a telescope. Societies often have telescopes that you can borrow, so gaining some hands-on experience, and at star parties you can see what each type of telescope (described in Chapter 3) has to offer. This will help you decide what is best for you.

History of astronomy

Observation, not experiment

In many sciences, a scientist sets up an experiment with the aim of testing an hypothesis, and from its results either proves or disproves the hypothesis – perhaps leading on to further experiments. In general, except on the surface of a planet like Mars, astronomers cannot carry out experiments. They have to observe what is happening in the universe and see if these observations comply with their theories. An observation may even give rise to a new theory. Hence, astronomy is based on observation, not experiment. In the following sections, we'll take a look at two examples of observational astronomy in action.

There was certainly no evidence of this!

The major problem associated with this theory, highlighted in the case of Mars, was that Aristotle and Ptolemy believed that planetary motions could only be in perfect circles. Mars is seen to move backwards (retrograde) during part of its motion across the sky (normally west to east). To accommodate this and other less obvious effects, Ptolemy theorized that the planets moved in small circles, called epicycles; so, for example, Mars was moving on an epicycle whose centre was itself moving around the deferent centred on the Earth.

Galileo Galilei's proof of the Copernican theory

Aristotle and Ptolemy believed that the Earth was at the centre of the universe and surrounded by concentric shells carrying the planets, the Sun and fixed stars. Mercury appeared to move the least across the sky, so perhaps it lay on the nearest shell, followed by Venus, the Sun, Mars, Jupiter and Saturn. Outside them all lay the fixed stars. We shouldn't, with hindsight, be too dismissive of these ideas. For, if this were not the case, the Earth would have to be rotating so that those on the surface were moving at up to 1000 miles per hour. (The speed is given by 25,000 miles/24 hours at the equator. Away from the equator multiply this by the cosine of the latitude.)

In contrast, Copernicus proposed a Sun-centred universe in which the planets rotated around the Sun. This, to a large extent, eliminated the need for epicycles because the retrograde motion of Mars could be explained by the fact that, when the Earth is nearest Mars, we are moving quickly past it 'on an inside track' and it appears from the Earth to be moving backwards. (The Copernican theory does not, as is often thought, totally eliminate the need for epicycles since the planets' orbits are ellipses, not circles.)

Right *Nicholas Copernicus.*
Above right *Galileo Galilei.*

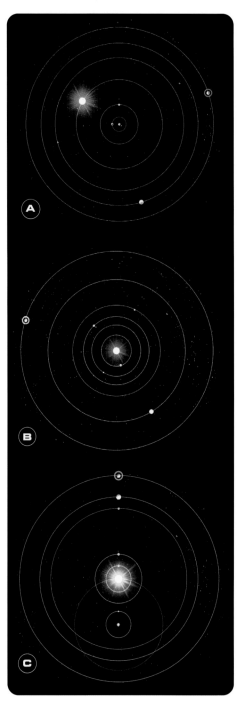

Galileo was able to prove that the Copernican theory was correct by making observations of the planet Venus with his small telescope. He deduced that, if the Ptolemaic theory was correct, Venus should move in an epicycle between the Earth and the Sun, with the centre of the epicycle rotating around the Earth at the same angular speed as the Sun. If so, Venus should remain reasonably close – so its apparent size should not alter too much – and, with the Sun always largely behind it, would always show thin crescent phases. Conversely, if the Copernican theory was correct then, when Venus was close to us (between the Earth and the Sun), it should appear relatively large and show thin crescent phases, but when it was visible towards the far side of the Sun it should appear smaller and show almost full phases. This was exactly what Galileo observed; Copernicus had to be right!

This proof of the Copernican theory was rein-forced by the fact that Jupiter was seen to be orbited by four moons, now called the Galilean Moons, showing that the Earth was not at the centre of all orbital motion. Within a year you could make these same observations, so why not make periodic observations of Venus even though there are no surface features to be seen?

Newton's Theory of Gravity

This theory was used to explain the motion of the Moon around the Earth and that of the planets around the Sun. Obviously, if one is to attempt to produce a theory that can predict the motion of

Left *Theories of the solar system.*
A – Ptolemaic (Earth-centred).
B – Copernican (Sun-centred).
C – Tycho (fixed Earth, Sun and orbiting planets travelling around it).

planets in their orbits, you need experimental data against which it can be tested. These data were produced as a result of one of the most productive periods of observation in the history of astronomy. They were made under the guidance of Tycho Brahe at an observatory called Uraniborg, built on the Danish Island of Hven, which lies to the south between Denmark and Sweden. (It is now called Ven and is part of Sweden.)

Tycho Brahe, born in 1546, was the son of a Danish nobleman. As a young teenager he took a great interest in astronomy – even though he was supposed to be studying law – and, still just 16 years old, made his own observations of the positions of the planets. He showed that the tables which predicted the motions of the planets were in error and thought that he could do better – it turned out that he was right! In 1572, he observed a 'new star' in the constellation of Cassiopeia – we would now call this a supernova. His observations showed that the 'star' really did belong to the firmament (the realm of distant stars) and was not just a local phenomenon within our solar system. It has become known as Tycho's supernova. Following these observations, the King of Denmark helped him found the observatory on Hven where, from 1577–1597, he undertook 20 years of meticulous observations of the stars and planets.

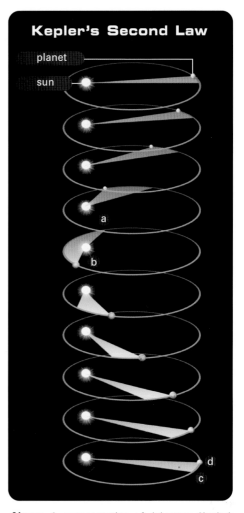

Kepler's Second Law

planet

sun

Above *A representation of Johannes Kepler's second law of planetary motion around the Sun.*
Above left *Isaac Newton (top) and Tycho Brahe.*

This was before telescopes were invented, but using sighting instruments which could measure the angle of elevation of a star as it crossed the meridian (the north-south line) and the time at which this happened, he was able to makea catalogue of stellar positions 10 times more accurate than had ever been achieved before.

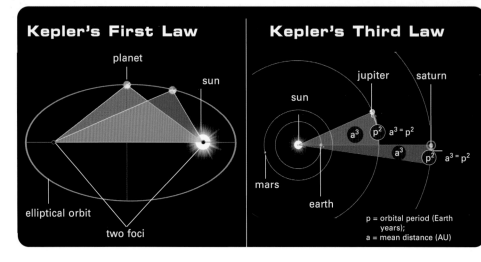

Kepler's First Law

planet

sun

elliptical orbit

two foci

Kepler's Third Law

jupiter   saturn

sun

a^3 p^2 $a^3 = p^2$

a^3

p^2 $a^3 = p^2$

mars

earth

p = orbital period (Earth years);
a = mean distance (AU)

More importantly, during this period he charted the movements of the planets across the heavens – data of immense value. He left Hven in 1597 and, in 1599, became the Imperial Mathematician to Emperor Rudolf II in Prague. There, in 1600, he was joined by Johannes Kepler to whom he gave the task of understanding the orbits of the planets – a task that took him many years.

When Tycho Brahe died in 1601, Johannes Kepler succeeded him as Imperial Mathematician. From the priceless data handed down to him by Tycho he was able to deduce his three laws of planetary motion (*see* p13 and illustrations above). The first two laws – planets orbit the Sun in elliptical orbits sweeping out equal areas in equal times – were published in 1609, and the third – the square of any planet's orbital period is proportional to the cube of its mean distance from the Sun, in 1619.

Isaac Newton, perhaps one of the greatest scientists of all time, was aware of these laws of planetary motion when he formulated his famous law of gravity. From his law – the force between two objects is directly proportional to the product of their masses divided by the square of the distance between them – one can easily deduce that the orbits of the planets should be ellipses, as stated by Kepler's first law, and that their periods and distances from the Sun should be related just as in Kepler's third law. This gave Newton the confidence to believe that his law, first deduced in relation to the Moon, did hold true over the vastly greater distances of the solar system; so it became known as Newton's Universal Law of Gravitation.

So you can see that accurate observations, made even with simple equipment, can lead to important results. A beginner would not expect to make new discoveries, but astronomy is a science where dedicated amateurs can make a real contribution to the subject.

Above left *Johannes Kepler used Tycho's planetary observations to derive the orbits of the planets, and published the Rudolphine Tables, which enabled the positions of the planets in the sky to be computed.*

A final word of encouragement

One of the great things about astronomy as a hobby is that the skies are always changing: the aspect of Saturn's rings varies over time, the prominence of the Great Red Spot on Jupiter waxes and wanes, meteor showers have their peaks and troughs of activity and the occasional lunar eclipse shows the beauty of a reddish-ochre coloured Moon. These things can be observed from year to year. But what can really make the hobby special are those astronomical events that are really rare such as solar eclipses, a transit of Venus or the passing of a bright comet. In July 1994, lumps of the comet Shoemaker-Levy 9 crashed into the surface of Jupiter and amateur astronomers were able to see for themselves the giant scars of debris on the surface – none could fail to be anything but transfixed at the sight!

There are rare times when the sky is so clear and atmosphere so steady, the fact that you are observing through a telescope is lost, one almost feels that one has left the Earth and is viewing from space! We live in a dynamic and exciting universe – amateur astronomy can make you feel part of it.

This book will give you nearly all the information that you would need to take up this hobby. It would be difficult to give precise details for observing the planets; for example, when Mercury is best seen. This type of information varies from year to year and one could never include information about events like the passage of a bright comet that may have only been discovered some months before. So this type of information needs to be found elsewhere. If you have access to the Internet, there are websites that will tell you each month what planets will be visible and where to look for them; as well as give you warning of exciting celestial events like the rare transit of Venus across the face of the Sun. Two such sites are, for the Northern and Southern hemispheres respectively,

1. http://www.jb.man.ac.uk/public/nightsky.html
2. http://homepages.win.co.nz/creation/ astronomy.html

Even if you can get the basic information from the Internet, it is worthwhile buying a monthly magazine that will give you detailed observing information and provide articles that will gradually increase your understanding of what you see. There are many prime magazines for amateur astronomers sold across the world in different languages.

A useful accessory to buy is a Planisphere (*see* right), which will let you know what you could observe at any time of the night – most useful when planning an evening or night's observing. Another useful aid to planning your observing is a first-class sky atlas.

Right *A refracting telescope allows you to see the Moon's mountains and lava-filled impact craters, and the ejecta lying around them.*

2

STARS AND GALAXIES

Were we to go far out into space and look back we would see a flat disc, visible by the light from myriad stars. Most of the stars would appear to lie along swirling spiral arms seemingly attached to the centre where there is a spherical bulge that we call the nucleus. This is our galaxy. We call it the Milky Way Galaxy from the appearance of the faint band of light that we see arching across the sky from Earth – our view of the Galaxy as seen from the Sun's location in a spiral arm about two-thirds of the way out from the centre. Many of the types of objects that can be seen within our galaxy, and the variety of other galaxies beyond will be described.

Opposite *The Milky Way Galaxy appears like a pale band of light arching across the sky on a dark night. The 'gaps' seen in the band are caused by dark dust clouds that lie in the spiral arms.*

STARS

How they are named

Virtually all the light by which we can see our galaxy comes from stars, the brighter of which make up chance patterns in the sky which the ancients grouped into constellations with names such as Cygnus (the Swan) and Taurus (the Bull). There are now 88 constellations that cover the sky whose boundaries and definitions were agreed in 1922 by the International Astronomical Union. Within each constellation the visible stars are named by Greek letters (*see* p25), called the Bayer letters, usually in order of brightness with alpha (α) the brightest, beta (β) the second brightest, and so on.

Within a constellation, Flamsteed numbers are also used in the naming of stars. In the Flamsteed catalogue, stars are numbered within each constellation from west to east to the limit of naked-eye visibility (51 Pegasai in the constellation Pegasus is an A-List example).

Brightness of stars

If we look up at the night sky we can discern two things about the stars that we observe, their brightness and, in some cases, their colour. The brightness alone doesn't actually tell us much about the star, as how bright it appears in the sky depends not only on how much light it emits but also on its distance from us – if two stars emit the same amount of light, the more distant one will obviously look fainter. If we can measure how far away a star

is from us, we can calculate how much energy it is emitting – called its luminosity. This is found to vary greatly, with the very brightest stars more than fifty thousand times brighter than our Sun and the faintest being about one-hundredth as bright. With our eyes alone we can only detect the colour of the brightest stars and can tell that Betelgeuse in the constellation Orion has a distinctly orange-red colour as does Aldebaran in Taurus. Capella in Auriga looks distinctly yellow while Rigel, also in Orion, is white-blue in colour. The colour is telling us about the surface temperature of the star.

A reddish star like Betelgeuse has a relatively cool surface temperature of 3500K, a yellow star like our Sun and Capella is about 6000K, while a blue-white star like Rigel has a surface temperature of some 11,000K.

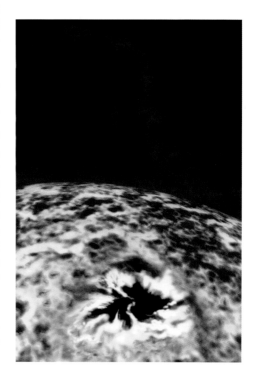

Right The Sun is Earth's own star. In comparison with other stars, our Sun is of average brightness, temperature and size.

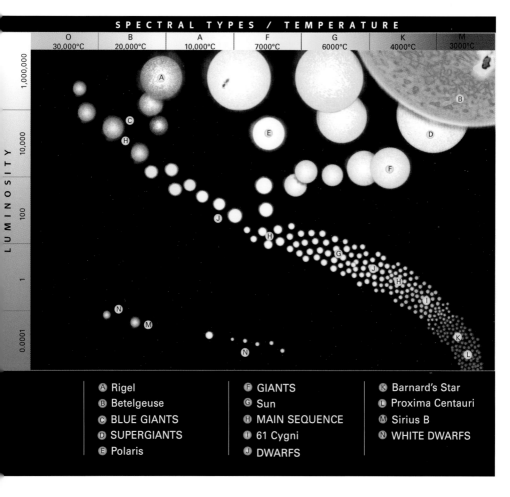

| O 30,000°C | B 20,000°C | A 10,000°C | F 7000°C | G 6000°C | K 4000°C | M 3000°C |

LUMINOSITY

1,000,000
10,000
100
1
0.0001

- Ⓐ Rigel
- Ⓑ Betelgeuse
- Ⓒ BLUE GIANTS
- Ⓓ SUPERGIANTS
- Ⓔ Polaris
- Ⓕ GIANTS
- Ⓖ Sun
- Ⓗ MAIN SEQUENCE
- Ⓘ 61 Cygni
- Ⓙ DWARFS
- Ⓚ Barnard's Star
- Ⓛ Proxima Centauri
- Ⓜ Sirius B
- Ⓝ WHITE DWARFS

A nuclear furnace

The brightness and colour of the surface of a star depend on how much energy is reaching it from the interior of the star. This energy is released by nuclear fusion converting light elements into heavier ones in the core of the star. Stars are formed as clouds of gas and dust contract down under gravity. As the gas is being compressed, its temperature rises and when it reaches around 10 million degrees at the centre, nuclear reactions can begin to generate energy – a star is born. The pressure produced by the nuclear furnace is sufficient to balance the gravitational forces trying to compress the star and it settles down into a stable state.

Above *The Hertzsprung-Russell diagram shows how a star's luminosity is plotted against its temperature.*

1. Star formation begins with clouds of gas and dust.

2. If disturbed, clouds collapse inward, and smaller, denser clumps separate from surrounding mass.

3. Denser clumps exert a gravitational pull, attracting more gas and dust.

4. Temperature and pressure increase, starting a nuclear reaction. A protostar is created.

11. The core of a massive star may collapse under such a powerful inner gravitational force that not even light can escape. This is a black hole.

10. The core collapses to a neutron star, which may be seen as a pulsar.

9. In massive stars (more than 8 times solar mass), nuclear fusion continues until a mighty explosion – a supernova – occurs.

8. The cooling, ageing star has reached its red giant phase.

5 Gas and dust clouds spin around the denser central protostar, collapsing to form a flattened disk.

6 A star (the Sun is depicted here) remains in a stable phase for most of its life, burning hydrogen into helium.

Once the star has depleted its hydrogen, it expands **7** and cools.

The life cycle of stars

Initially, and for most of their lives, stars convert hydrogen into helium, but in the latter stages of their life helium is converted into carbon, oxygen and other heavier elements. At this point the stars expand – our own Sun may even encompass the Earth – and their surface cools, so becoming orange or red in colour. As they are so big, they are very luminous and are called 'red giant' stars; Aldebaran in Taurus is one of these. The most massive stars form red super-giants such as Betelgeuse in Orion whose diameter is over half the diameter of the orbit of Jupiter! At the very end of their lives, stars ex-plode and the elements that have been created within them are ejected into space and form dust clouds which can impede our view of the galaxy beyond (as in the case of the Coal Sack in Crux – an A-List object; *see* p138). In the most massive stars, nuclear fusion can create elements up to iron in the periodic table (*see* Glossary), but in the immense explosions called supernovae (such as the A-List object, the Crab nebula in Taurus) that mark the end of their lives, even heavier elements are created such as lead and gold and uranium.

Over billions of years the space between the stars becomes enriched with heavier elements, so that when new stars are born there may be sufficient material out of which planets like the Earth can form and on which life might arise. The atoms that make up much of our bodies and the air we breathe were all produced in stars!

Left *Star birth and death is a cyclical process where the leftover gas and dust from a previous generation of stars become the building mater-ial for the next generation.*

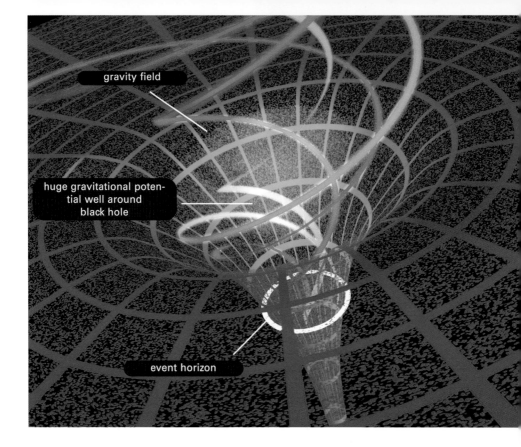

gravity field

huge gravitational poten-
tial well around
black hole

event horizon

Pulsars and black holes

Stars leave behind remnants derived from their cores; stars about the mass of our Sun leave behind a small 'white dwarf' star about the size of our Earth. These are found at the centre of planetary nebulae such as the A-List objects, the Ring Nebula in Lyra and the Dumbbell Nebula in Vulpecula. More massive stars produce neutron stars – no bigger than a city but weighing more than our Sun. Sometimes these radiate pulses of light and radio waves as they rotate – rather like interstellar lighthouses – and can be detected as pulsars. There is one at the heart of the Crab Nebula supernova remnant mentioned earlier. The cores of the most massive stars collapse so far under gravity that they become black holes: regions of space with so much gravitational pull that not even light can escape!

Luminosity and mass

It turns out that the luminosity of a star depends on its mass. The more massive the star, the hotter its core becomes and the faster it converts hydrogen into helium. More energy reaches the surface, increasing its temperature

Above *An artist's impression of the formation of a black hole.*

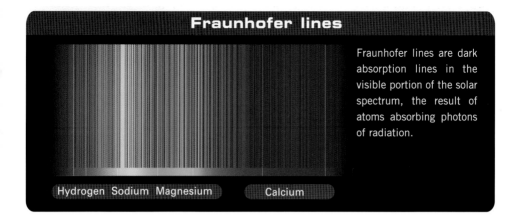

Fraunhofer lines

Fraunhofer lines are dark absorption lines in the visible portion of the solar spectrum, the result of atoms absorbing photons of radiation.

Hydrogen Sodium Magnesium   Calcium

and making it emit more heat and light. A massive star may be 10 times the mass of our Sun, but it emits perhaps a 1000 times more energy. The amount of hydrogen that can be fused is about 10% of the total mass. Thus the massive star will burn up its fuel at a rate 1000 times faster than our Sun, but only has 10 times more

Above *The Fraunhofer lines provide crucial information about the chemical composition of the Sun.*

fuel, so its life will be one-hundredth as long. Massive stars thus have short lives in comparison to our Sun which will live for around 10 billion years in total. These massive stars with very high surface temperatures – appearing blue-white to us – give off ultraviolet light which is able to excite the surrounding gas and make it glow. They do not live for long so are found in regions of recent star formation. These, like the A-List object, M42 in Orion, thus stand out in photographic images due to the beautiful pink-red glow of excited hydrogen.

THE GREEK ALPHABET

α	alpha	ν	nu
β	beta	ξ	xi
γ	gamma	ο	omicron
δ	delta	π	pi
ε	epsilon	ρ	rho
ζ	zeta	σ	sigma
η	eta	τ	tau
θ	theta	υ	upsilon
ι	iota	φ	phi
κ	kappa	χ	chi
λ	lambda	ψ	psi
μ	mu	ω	omega

In contrast, stars with low masses compared to our Sun burn their hydrogen very slowly and have lifetimes longer than the present age of the universe (about 14 billion years). They glow faintly with reddish light so are hard to see at great distances from us.

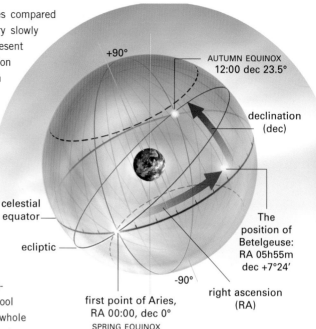

The classification of stars

The visible surface of the Sun or any star is known as the photosphere. Radiation produced deep within the star passes through this cool outer layer which contains the whole range of elements out of which the star was formed. Their effect is to absorb light at specific frequencies (or colours) so that, if the light of the star is split up into the different colours of the spectrum, dark lines are seen where specific elements have absorbed a particular colour of light. These are called Fraunhöfer lines (see p25) after Joseph Fraunhöfer who observed them in the spectrum of the Sun and correctly deduced their cause. The pattern that we see for a particular star is highly dependent on its surface temperature, so this has given astronomers a way of classifying stars according to their spectrum.

The current classification system – developed at Harvard Observatory – classifies stars by a series of letters in order of highest temperature to lowest. The seven main types are O, B, A, F, G, K, M (see H-R diagram on p21). Each class is subdivided into subclasses such as G0 to G9, with G0 being the hottest and G9 the coolest within the class. Our Sun is a star of type G2.

MEMORIZING SPECTRAL TYPES
The spectral types O, B, A, F, G, K, M are known to all astronomy students by the mnemonic 'Oh Be A Fine Girl Kiss Me'.

Charting the stars

Since ancient times the positions of stars have been plotted on spheres or star charts – often accompanied by beautiful drawings of the constellation figures. This requires a suitable coordinate system in just the same way as we locate cities on the surface of the Earth by specifying their latitude and longitude. In the case of the stars, we locate them on a 'celestial sphere' centered on the Earth. The projection of the Earth's north and south poles cut the celestial sphere at the north and south celestial poles respectively, while the projection of the Earth's equator onto the celestial sphere defines the celestial equator. The 'Declination' (Dec) of a star – the equivalent to latitude – is measured in degrees north and south of the celestial equator with the north and south celestial poles having declinations of +90° and -90° respectively. The second coordinate – comparable to longitude – is called 'Right Ascension' (RA) and is measured eastward around the celestial equator. In just the same way that there has to be an arbitrary zero of longitude – the Greenwich Meridian – we need some zero of right ascension. It is called the 'first point in Aries' and is the point where the ecliptic (the apparent track of the Sun across the sky) crosses the celestial equator northwards at the vernal equinox, so marking the first day of spring. One might well assume that this point will be found in the constellation Aries, but if you locate it on a star chart you may be somewhat surprised to see that it is now in the constellation Pisces. This is due to the fact that the Earth's rotation axis precesses – just like a top – but in this case it takes some 26,000 years to make a complete revolution, and therefore is not so obvious to us. But it does mean that over time star positions will change and, for example, after a few hundred years or so Polaris will no longer be a good 'pole star'. This is why star charts have a date – termed their 'epoch' with current star charts being of epoch 2000.

RA is usually measured in hours, minutes and seconds from 0:00 to 24:00. One hour of RA, the distance the Earth rotates in an hour of time, is equal to 15° at the equator.

Opposite top An imaginary sphere helps astronomers map the skies. The projection of Earth's equator and two poles onto the sphere marks the celestial equator, and celestial north and south poles.

Above *The Sun (our own star) in ultraviolet light photographed by the SOHO spacecraft showing plumes of hot gases arcing along the Sun's magnetic field lines.*

Bear in mind that:
a complete circle = 360°
1° = 60 arc minutes
1 arc minute (1') = 60 arc seconds (60")

Trying to estimate the number of degrees between features in the night sky for the first time can prove a daunting task. However, using your hand can prove an effective measuring scale. For example, your raised index finger held at arm's length in front of your face marks off about 1°; your outstretched hand is about 20°, which will cover the distance across the bright stars of the Plough, or Big Dipper. Both the Sun and the Moon have an angular diameter of about 0.5°.

A STAR'S MAGNITUDE

On a star chart, stars are represented by filled circles whose size increases for brighter stars, so giving some impression of how the sky might appear. Ancient Greek astronomers such as Hipparchus, measured the brightness of individual stars by grading them according to a scale of magnitude from 1 (the brightest) to 6 (the faintest that could be seen with the unaided eye). In the 19th century, Norman Pogson refined the scale by giving it a mathematical basis. Measurements showed that a 1st magnitude star was about 100 times brighter than a 6th magnitude star, and he realized that the magnitude scale as measured by the eye corresponded to a logarithmic scale of intensity. He thus defined a 5-magnitude difference to be exactly 100 which, using a logarithmic scale, made 1-magnitude difference a factor of 2.512 times in brightness.

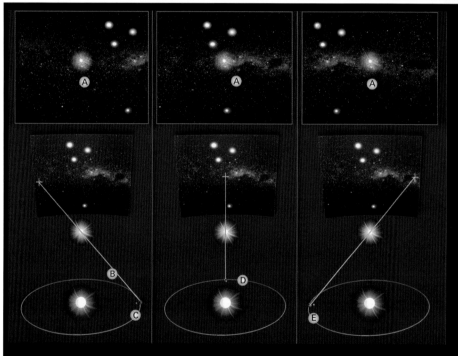

A A star as it is observed against the night sky in relation to a small group of stars.

B C Line of sight (B) from the position (C) of the Earth in its orbit around the Sun. The star appears to lie to the left of the group of stars.

C Earth has moved along its orbit so that the star being observed lies just below the group of comparison stars.

E Six months from the start of the observations, Earth is at the opposite extreme of its orbit and the star appears to the right of the comparison stars.

The distances of stars: stellar parallax

In order to determine the luminosity of stars, we need to find their distances. This is not an easy task as they are very far away. However, if we study a small patch of sky (very) carefully as the Earth travels around the Sun, any star that is nearby will appear to move in a tiny ellipse relative to the positions of distant stars, and return to the same position in the sky in a year's time. This is called 'parallax', and is used to measure distances to the relatively few stars close enough to the Sun to show a measurable parallax. The parallax of even the nearest star is extremely small.

Astronomers measure distances in parsecs, where 1 parsec is the distance of a star whose parallax is 1 arc second, i.e. 1/3600th of a degree (although no star is close enough to have a parallax of 1 arcsec). One parsec equals the distance that light can travel in 3.26 years so it is equivalent to 3.26 light years. The closest star to the Sun is Proxima Centauri which has a parallax of 0.76 arcsec, corresponding to:

$$d = {}^1/_p = 1.3 \text{ parsec}$$

(where distance is d and parallax is p)

Opposite *A representation of stellar parallax, which is used to measure the distances to stars.*

Spectroscopic parallax

The distances of stars beyond those whose direct parallax can be measured is done by an indirect method based on their spectral type – hence the term 'spectroscopic parallax'. We make the assumption that a star of type, say, F5 whose parallax and hence distance is known will have just the same brightness as a more distant star, also of type F5, whose parallax cannot be measured. If this assumption is correct, we can easily determine the distance of the more distant star.

Suppose the nearby star is 4 parsecs away and the distant star appears 100 times (5 magnitudes) less bright. From the inverse square law we know that this star will be 10 times further away (the square root of 100) and thus at a distance of 40 parsecs. This method allows distances across the Galaxy to be measured with the proviso that the apparent brightness could be reduced due to the presence of dust between a distant star and us.

The fundamental distance unit for these measurements in the solar system is the mean distance of the Earth from the Sun called the 'Astronomical Unit', and so this forms the bedrock upon which all distance measurements across the universe are made where 1 AU = 149,597,870km (92,584,300 miles)

1 light second (distance light travels in a second) = 300,000km (186,000 miles)

1 light minute (distance light travels in a minute) = 18 million km (11 million miles)

1 light year (distance light travels in a year) = 9.4 million million km (6 million million miles)

Light curve of eclipsing binaries

time

brightness

(A) Various stages of a binary star's orbit.

(B) Perspective of an eclipsing binary as viewed from Earth.

(C) Light curve falls when brighter star is eclipsed by dimmer star.

(D) Light curve dips slightly when dimmer star is eclipsed by brighter star.

APPARENT AND ABSOLUTE BRIGHTNESS OF STARS

The brightness of the stars seen by eye and plotted on star charts is called the 'apparent brightness' and depends on both the actual brightness of a star and its distance from us. If all the stars could be lined up at a fixed distance from Earth, then those stars that appear brighter than others really would be brighter and vice versa. So, in our mind's eye, we line the stars up at a distance of 10 parsecs (36.2 light years) from us. The brightness that each star would then have is called its 'absolute brightness'. If an observed star were closer than 10 parsecs, its absolute brightness would be less than its apparent brightness as, in our mind, it has been moved further away from us while, if it lay beyond 10 parsecs, it would need to be brought closer to us and its absolute brightness would be greater than its apparent brightness.

Multiple star systems

On a detailed star chart, stars are sometimes shown as overlapping. This indicates that they are binary or multiple star systems. Occasionally this is simply because two stars at different distances from us just happen to appear very close together in space, and are then called 'optical doubles'. More usually, double stars are pairs of gravitationally bound stars that orbit each other around a common centre of gravity. They are extremely common in the Galaxy and surveys have shown that about half of all sun-like stars have one or more companions. The A-List star Albireo in Cygnus is one such binary system. Sometimes in what initially appears to be two stars, one (or more) of them is itself shown composed of two stars, so we have a multiple star system. The A-List star system Epsilon Lyrae, the double-double in Lyra, has a total of four stars with two pairs of stars in orbit around each other, while the A-list star system Castor in Gemini contains six stars.

Eclipsing binaries

The stars in many binary star systems are too close to each other to be seen separately but if their orbit is aligned to our advantage, at times one of the stars can pass in front of the other causing the brightness of the pair to drop for a while. A typical light curve of these 'eclipsing binaries' (*see* opposite page) shows a near constant glow for much of the cycle, with one or two clear dips as one star passes in front of the other. The shape of the dips – whether they are of short duration, deep, or wide and shallow – can tell us much about the types of stars involved. The A-List star Algol in Perseus is an excellent example of an eclipsing binary.

Spectroscopic binaries

The fact that a star is actually a binary system, even when both stars cannot be observed as such, can also be deduced by observing its spectrum. The motion of one star around the other produces a Doppler shift (*see* Glossary) in the spectral lines. Such star pairs are called spectroscopic binaries as this is how they are detected. The three 'stars' that make up the Castor system mentioned earlier are each spectroscopic doubles.

Variable stars

Some stars are shown on a star chart as a filled circle surrounded by a ring. This indicates that the star is observed to change in brightness. Such stars are called 'variable stars'. The brightness of such a star can be plotted as a function of time to give what is called a 'light curve'. The shape of the light curve is important in distinguishing the underlying cause of the variability. Some stars vary with a regular

repeating cycle: these are periodic variables. Other stars are much less predictable and may flare up or drop in brightness with little or no warning. There are even stars showing both types of activity. A star may also vary in brightness if it is a binary or multiple star system when, as in the case of Algol mentioned earlier, the light output may change as the stars alternately become eclipsed.

Intrinsic variables

As stars are in the final stage of contraction, they pass through the so-called T-Tauri phase – this is when the light output can be extremely variable. Eventually they settle down and spend most of their lives in a more-or-less stable state but when, after all the hydrogen has been converted into helium and reactions to form heavier elements are in progress, they can become strongly variable again as they come to the end of their lives.

Pulsating Cepheid variables

Cepheid variable stars are very luminous and have a regular pattern of light variation; first increasing to maximum brightness and then, more slowly, falling back to minimum brightness. The time over which this cycle repeats is very stable. These are named after their prototype star, Delta (δ) Cephei, but the first to have been discovered was actually the A-List star Eta Aquila. They are some of the very brightest stars so can be seen at great distances. The period of their brightness variations has been shown to be proportional to their luminosity, so that if the peak brightness of a Cepheid variable star observed in a distant galaxy is

On a dark night when the skies are transparent, the band of light that we can see arching across the heavens shows that we live toward the edge of the Milky Way Galaxy, or simply 'the Galaxy'. The light comes from the myriad stars packed so closely together that our eyes fail to resolve them into individual points of light. The band of light is not uniform; the brightness and extent is greatest nearing the constellation Sagittarius, suggesting that in this direction we are looking toward the galactic centre. Dust ejected from stars at the end of their life prevents us from seeing more than one-tenth of the way towards the centre in visible light. However, in infrared light, we can 'see' right to the centre. In the opposite direction in the sky, the Milky Way is less apparent – implying that we live toward one side. Also, the fact that we see a band of light tells us that the stars, gas and dust that make up the Galaxy are in the form of a flat disc. The major 'visible' constituent of the Galaxy, about 96%, is made up of stars, with the remaining 4% split between gas (about 3%) and dust (about 1%). Here, 'visible' means that we can detect the constituents by electromagnetic radiation. As discussed later, astronomers suspect there is a further component of the Galaxy we cannot detect, called 'dark matter'.

measured together with its period, it is possible to calculate its distance. This is how the distances to galaxies across the universe have been determined, so Cepheid variable stars are of very great importance.

Cataclysmic variables

As their name suggests, these stars vary suddenly and unpredictably. They are close binary systems in which, initially, one star has ended its life as a white dwarf orbiting a normal star. As the companion star evolves to become a red giant, its outer atmosphere swells to a size such that the gravitational pull of the white dwarf on its outer layers can pull material onto the white dwarf causing an eruption. The uneven flow of material from the red giant onto the white dwarf gives rise to flare-ups or explosions of varying duration.

Above *An artist's rendering of a cool orange and hot white star-pair of an eclipsing binary system.*

Star clusters

Against the background of the Milky Way and the general backdrop of stars one can pick out concentrations of stars. These we call 'clusters', of which there are two distinct types.

- **Open clusters** are groups of young stars found in the plane of the Galaxy. A well-known example (and one of the youngest clusters) is the A-List object, the Pleiades. The brightest stars of the cluster are luminous blue 'B-type' stars which are massive stars in the hydrogen-burning phase of their evolution. These form a tight group in the night sky easily seen with the naked eye in the constellation Taurus.

  It is virtually impossible for a single star to form in isolation, so stars always form in groups of hundreds to thousands of stars. Over time the stars will tend to drift apart but, while they are young, the stars are relatively closely packed together and form the open clusters that we see. Several more are A-List objects.

- **Globular clusters** are spherical in shape and contain several tens of thousands to perhaps a million stars. They are systems of very old stars, formed at about the same time as the Galaxy itself and are found surrounding the galaxy, in what is called the galactic halo, as well as in the central bulge. About 150 globular clusters are known, of which the A-List objects Omega (ω) Centauri – visible as a fuzzy patch in the Southern sky – and M13 in Hercules are two of the best examples.

Below *A double cluster, NGC 1850, in the Large Magellanic Cloud: surrounding the stars is a mass of diffuse gas which scientists believe has been created by the explosion of massive stars.*

Nebulae and the interstellar medium

Together the gas and dust between the stars make up what is called the 'interstellar medium' (ISM). Most of this is not apparent to your eyes, but in some regions you can see either 'emission nebulae', where the gas is glowing, or 'dark nebulae', where a dust cloud appears in silhouette against a bright region of the Galaxy. One of the most spectacular examples of an emission nebula is the Great Nebula in Orion (an A-List object, M42), a region of star formation where the hydrogen gas is being excited by ultraviolet light emitted by the very hot stars, forming the Trapezium at its heart.

An example of a dark nebula is the Coal Sack, seen against the background of the Milky Way close to the Southern Cross. Sometimes we can detect dust due to the fact that it scatters

Above In the nebula NGC 3603, an entire stellar life cycle is depicted, starting with the giant gaseous pillars hiding embryonic stars, followed by young stars encircled with disks, massive young stars in the starburst cluster. The blue supergiant (upper left of centre) marks the end of the life cycle.

light from nearby stars. Called a 'reflection nebula', an excellent example lies close to the star Merope in the Pleiades. This appears blue in colour, partly because the light from this very hot star is bluish anyway, but also because the dust scatters blue light more readily than red light (the same reason that our skies are blue).

Size and structure of the galaxy

The size of the Galaxy was first measured by Harlow Shapley, who used Cepheid variables to measure the distances to 100 of the globular clusters associated with the Milky Way. He found that they formed a spherical distribution, whose centre should logically be that of the Galaxy, and deduced that the Sun was about 30,000 light years distant. The Galaxy diameter was about 100,000 light years across. The most common element in the universe, hydrogen, has given us a way to study the spiral structure of the galaxy. Neutral hydrogen (HI) emits a radio spectral line with a specific wavelength. Radio observations of this line along the plane of the Milky Way show that the gas in the disc is concentrated into clouds. Determining their velocity using the Doppler shift, these data are used to plot out the positions of the gas clouds; a pattern of spiral arms emerges, indicating that we live in a typical spiral galaxy. It is now believed that the Sun is 28,000 light years from the galactic centre and, using spectroscopic measurements to observe its motion relative to the globular clusters (essentially fixed in space), we can calculate that the Sun is moving around the centre of the Galaxy at about 200km/sec (125 miles/sec), taking approximately 250 million Earth years to travel once around it. It appears that the central parts of the Galaxy rotate like a

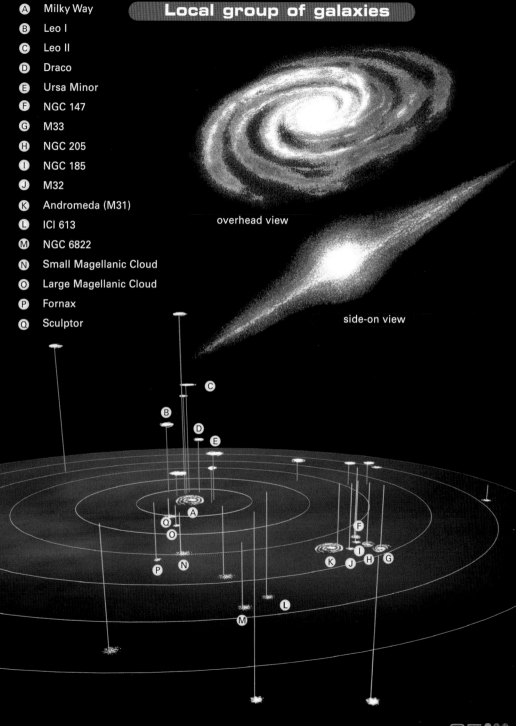

Local group of galaxies

A — Milky Way
B — Leo I
C — Leo II
D — Draco
E — Ursa Minor
F — NGC 147
G — M33
H — NGC 205
I — NGC 185
J — M32
K — Andromeda (M31)
L — ICI 613
M — NGC 6822
N — Small Magellanic Cloud
O — Large Magellanic Cloud
P — Fornax
Q — Sculptor

overhead view

side-on view

solid body so that the rotation speed increases as one moves out from the centre. Further out, the rotation slows so that stars have longer 'galactic years'; a plot of rotation speed against distance from the centre produces what is called the 'galactic rotation curve'.

Galactic rotation curve

The way in which the stars rotate around the centre of our galaxy has proved difficult to understand due to two significant problems. The first is related to the spiral structure. In its life, the Sun has circled the galactic centre about 20 times, so why haven't the spiral arms wound up? The solution is hinted at by a visual clue. Spiral arms seen in other galaxies stand out because they contain many bright blue stars – and bear in mind that a single, very hot star can outshine 50,000 suns like ours. Because these stars have very short lives they must be young, so the spiral structure seen now is not that which would have been observed in the past. As first suggested by Swedish astronomer Bertil Lindblad, it appears that the spiral arms are transitory, and caused by a spiral density-wave rotating round the galactic centre – a ripple that sweeps around the Galaxy, moving through the dust and gas. This ripple compresses the gas as it passes and can trigger the collapse of gas clouds, so forming the massive blue stars that delineate the spiral arms. The young blue stars show us where the density wave has just passed through, but in its wake will be left myriad longer-lived (less bright) stars that form a more uniform disc.

Galactic dark matter

The second problem is that we would expect the rotation speeds of the stars further away from the centre to fall off with distance, in the same way that the speeds of the planets in the solar system do. This is based on the supposition that the gravitational effects on the outer stars are produced solely by the mass that we can see. But in fact, the rotation speeds tend to remain relatively constant giving a flat, rather than declining, rotation curve. One can only reconcile the observed motions of the outer stars with the hypothesis that the whole Galaxy lies within a halo of so-called 'dark matter' (it gives off no radiation by which we can detect it) whose total mass is several times greater than the visible matter within the Galaxy. Although ideas exist of what might make up this dark matter – physicists have suggested exotic particles such as axions (neutral elementary particles) and WIMPs (Weakly Interacting Massive Particles) – none has yet been detected, though searches are actively underway.

Other galaxies

Galaxies, originally called 'white nebulae', have been observed for hundreds of years, but it was not until the early part of the last century that the debate as to whether they were within or beyond our Galaxy was settled. This essentially became possible when observations of Cepheid variables enabled their distances to be measured. These nebulae are, of course, objects outside our Galaxy and are now observed throughout the universe.

Galaxies can be divided into a number of types and then subdivided further within a classification scheme first devised by Edwin Hubble. As more and more galaxies were discovered, it became apparent that they cluster into groups containing tens to thousands of galaxies having differing forms and masses.

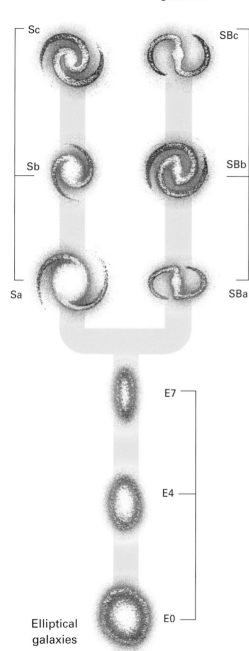

Spiral galaxies

Barred spiral galaxies

Sc

Sb

Sa

SBc

SBb

SBa

E7

E4

E0

Elliptical galaxies

HUBBLE'S 'TUNING FORK' CLASSIFICA-TION OF GALAXIES:

Type E0 – M87
Giant spherical galaxy at the heart of the Virgo Cluster (*see* A-List); a powerful source of radio emission and there appears to be a massive black hole at its core (now believed to be the case with all galaxies).

Type E4 – M49
Another, more elongated giant elliptical galaxy in the Virgo Cluster; made up of Population II stars, star formation has largely ceased.

Type E6 – M110
A highly elongated elliptical galaxy accompanying M31.

Type Sa – M104
The Sombrero Galaxy in the A-List has a very prominent nucleus surrounded by very tightly wound spiral arms.

Type Sb – M31
The Andromeda Galaxy (*see* A-List) has a medium-sized nucleus with tight spiral arms around it.

Type Sc – M51
The Whirlpool Galaxy in the A-List has a small nucleus and open spiral arms.

Type SBa – M83
Lying in Hydra, M83's tight spiral arms extend from a bar across the prominent nucleus.

Type SBb – M95
Situated in Leo, this A-List galaxy has a very prominent bar extending out from the nucleus.

Type Sc – M109
Situated in Ursa Major, M109 has open spiral arms and a small nucleus with an extended bar.

Types of galaxies

Elliptical galaxies

These have an ellipsoidal form rather like a rugby ball. They range from those that are virtually circular in observed shape, called EO by Hubble, to those, called E7, which are highly elongated. One interesting fact is that ellipsoidal galaxies do not appear to have any young stars within them. Star formation appears to have ceased, all the gas having been used up to form stars in the past.

At the heart of large galaxy clusters, one or more giant elliptical galaxies are often observed, some nine times the mass of our own Galaxy. They are probably the result of the merger of many smaller galaxies; these are the most massive of all galaxies but are comparatively rare. Far more common are elliptical galaxies containing perhaps a few million solar masses within a volume a few thousand light years across.

Spiral galaxies

Like our own Galaxy, these have a flattened spiral structure. They make up the majority of the brighter galaxies. Edwin Hubble classified them first into four types: S0, Sa, Sb and Sc. S0 galaxies have a very large nucleus with hardly visible, very tightly wound, spiral arms. As you progress toward type Sc, the nucleus becomes relatively smaller and the arms more open. In many galaxies the spiral arms appear to extend from either end of a central bar. These are called 'barred spirals' and are delineated SBa, SBb and SBc.

Irregular galaxies

A small percentage of galaxies show no obvious form and are classified as 'irregulars'. One nearby example is the Small Magellanic Cloud (SMC). Its companion, the Large Magellanic Cloud (LMC), is usually classified as one too, though it shows some features of a small barred spiral. Such small galaxies are not very bright so we cannot see too many, but they may, in fact, be the most common type. They contain enough gas to allow star formation, but relatively less dust than found in our Galaxy. Both the LMC and SMC are in the A-List. The LMC contains one of the largest regions of star formation visible to us. This is also an A-List object, called the 30 Doradus or Tarantula nebula in Dorado.

Left *A blue ring dominated by clusters of hot, young, massive stars creates a pinwheel around a yellow galaxy of older stars – known as Hoag's Object.*
Opposite *The Whirlpool Galaxy (known also as M51 and NGC 5194) is a classic spiral galaxy.*

Starburst galaxies

These emit more than usual amounts of infrared light and radio waves. As a result they came to prominence when surveys of the sky were made with infrared telescopes. A nearby example is the A-List galaxy M82, 12 million light years away in the constellation Ursa Major (the Great Bear). It appears that a close passage with M81, its neighbour in space, has triggered a rapid burst of star formation. The radiation from young stars heats up the dust in the galaxy, so producing the infrared emission. Among the stars that are born will be

Below *The barred spiral galaxy NGC 1512 is filled with infant star clusters that extend 2400 light years across.*

a small number of very massive stars that evolve quickly and then die in spectacular supernova explosions.

Active galaxies

These are galaxies where some processes going on within them make them stand out from the normal run of galaxies, particularly in the amount of radio emission they produce. At the heart of our Galaxy lies one of the strongest radio sources in the Milky Way. However, Sgr A* would be too weak to be seen from a distant galaxy and our Galaxy would therefore be termed a 'normal' galaxy. However, some galaxies do discharge vastly more radio emission, shining like beacons across the universe – and because most of the excess emission lies in the radio part of the spectrum, these are called

radio galaxies. Others produce an excess of X-ray emission and, collectively, all are called active galaxies. Though relatively rare, there are obviously energetic processes going on within them that make them interesting objects for astronomers to study.

The brightness of active galaxies is often seen to vary on time scales measured in just hours. As nothing can vary more quickly than the time it takes light to travel across it, this means that the size of the emitting region cannot be more than a few light hours across – no bigger than our solar

system. It thus appears that the radiation is coming from a very small region at the heart of the galaxy that we call an Active Galactic Nucleus – or AGN.

Above *The composite image of the Sombrero Galaxy is one of the largest Hubble mosaics ever assembled, and was released to celebrate the Hubble Heritage team's five-year anniversary. Six photographs were stitched together to create the image.*

NOTE

Astronomers believe that supermassive black holes, containing up to several billion solar masses, exist at the centre of all large elliptical and spiral galaxies. In the great majority of galaxies these are quiescent, but in some, matter is currently falling into the black hole, fuelling the processes that give rise to the X-ray and radio emissions. The active galaxies M87 in Virgo and Centaurus A in Centaurus are both A-List objects.

Galaxy groups and clusters

Our Local Group

Most galaxies are found in groups typically containing about ten galaxies or clusters that may contain up to several thousand. The Milky Way Galaxy forms part of what's known as the Local Group (*see* p35), which contains around 40 galaxies within a volume of space that's three million light years across. Our Galaxy is one of the three spiral galaxies which, along with M31 in Andromeda and M33 in Triangulum (both A-List objects), dominate the group and contain the majority of its mass. M31 and our own Galaxy are comparable in size and mass, and their mutual gravitational attraction is bringing them toward each other so that in a few billion years they may well merge to form an elliptical galaxy. The group contains many dwarf elliptical galaxies and there are several large, irregular galaxies such as the Small and Large Magellanic Clouds with

at least 10 dwarf irregulars to add to the total. There may well be more galaxies within the group, hidden beyond our Milky Way which obscures over 20% of the heavens.

Superclusters

Small clusters and groups of galaxies appear to make up structures on an even larger scale. Known as superclusters, they have overall sizes of around 300 million light years (100 times the scale size of the Local Group). Usually a supercluster is dominated by one very rich cluster, surrounded by a number of smaller groups. The Local Supercluster is dominated by the Virgo Cluster (so called as its galaxies are seen in the direction of the constellation Virgo), which contains over 2000 galaxies. The Virgo Supercluster, as it's often called, is in the form of a flattened ellipse about 150 million light years in extent, with the Virgo Cluster at its centre and our Local Group near one end. The A-List chart for the constellation Virgo shows the region, called the

'realm of the galaxies' that lies towards the heart of our own supercluster. Such superclusters are found throughout the universe which has an overall structure rather like a sponge with the clusters of galaxies surrounding empty regions of space called voids; the result, we believe, of gravity working on the slight variations in density laid down in the Big Bang origin of the universe.

Owing to the working of the fundamental forces of matter on the hydrogen and helium that were formed in the Big Bang, many diverse and often beautiful objects have been created. It is hoped that the remaining chapters of this book will help you to find and observe them and so see some of the wonder of the universe yourself.

Above *Antares, Alpha (α) Scorpii, and Rho (ρ) Opiuchi illuminate surrounding gas and dust in the Milky Way over the Southern Hemisphere.*
Left *This spectacular scientific visualization of two galaxies colliding has been created by the Space Telescope Science Institute as part of a full-dome video sequence for the Smithsonian Institution.*
Opposite *The Coma Cluster, one of the densest galaxy clusters known, contains thousands of elliptical galaxies, each of which houses billions of stars.*

3

ASTRONOMICAL INSTRUMENTS

In using our eyes, binoculars or small telescopes to observe the heavens, we can derive great enjoyment and satisfaction – but we should not expect too much! We do not see colour – except for the planets and some bright stars – because of the low light levels and we cannot expect to see similar views to those taken by the Hubble Space Telescope! Personal telescopes will, however, show you rich detail on the Moon, the surface of Jupiter and its Galilean moons, Saturn and its rings, rich star clusters and, beyond our own Milky Way, other galaxies lying millions of light years away. The wonder of it is that a photon that's just registered on your retina may have left a distant galaxy many millions of years ago!

Opposite *If your telescope is to remain outdoors for an extended period of time, precautions should be taken to protect the lenses and mirrors from dew and dust.*

Your eyes

If you wish to observe the constellations, meteors or the Milky Way, your eyes are the only optical instruments for the purpose. To use them to view the heavens at their best, however, a little time is required for them to become adapted to the dark. There are two effects:

- In dark conditions the pupils dilate allowing more light to enter the eye. This happens over a period of 20 seconds or so, thus quickly enabling you to see fainter stars.
- The second mechanism (contained in the biochemistry of the eyes) takes about 20 minutes to come into effect. When high light levels are not reaching the retina, vitamin A is converted first into retinene, then into rhodopsin, which significantly improves the sensitivity of the retina's rods and cones.

Even after dark-adaptation, how much you see on any given night depends critically on the amount of dust and water vapour in the atmosphere. These absorb and scatter the starlight, so making it difficult to see fainter objects. The dust and vapour also scatter back toward your eyes any light from the ground – called light pollution – making the sky appear brighter and visibility worse. Try to get as far as possible from built up areas – even a few miles will help significantly. The term 'transparency' is used to define how clear the sky is. At a dark site with a very transparent sky, you'd typically be able to see a 6th-magnitude star if looking toward the zenith.

Binoculars and telescopes, however, enable you to see fainter objects since their objective lens collects more light than your eyes. They can also magnify the image to enable you to see more detail – for example, on the surface of the Moon or Jupiter.

TIPS FOR DARK-ADAPTING YOUR EYES

Use red light when reading sky charts so that your eyes remain adapted to the dark – you could place red cellophane paper over the end of your ordinary flashlight if it does not have a red filter.

By directing your gaze slightly to one side of the object of interest, its light will fall on the more sensitive outer parts of your retina making fainter objects become more visible. This technique is known as 'averted vision'.

Binoculars

A good pair of binoculars is a wonderful instrument for observing the heavens and every astronomer should have one. The two numbers that are used to specify a pair of binoculars, such as 8x40, are first, the magnification, and second, the aperture of the objective lens in millimetres. The bigger the aperture the more light

Right 25x100 binoculars need to be mounted on a tripod in order to view stars comfortably.

is collected, enabling fainter objects to be seen. As described in detail later, increasing the magnification also helps one to see fainter stars. However, the greater the magnification the smaller the area of sky that can be seen at one time. 8x40 and 10x50 are both very good for astronomical use.

Lens and prism coating

The amount of light that exits through the eyepiece depends on the quality of the coatings put on the lenses and prisms to minimize reflected light. If all glass surfaces are multicoated, then the binoculars will transmit almost all the light. If the coatings are nonexistent or less good, light is reflected internally and scatters into the image, reducing its contrast. This is one reason why really good binoculars are expensive.

A second reason relates to the complexity, and hence cost, of the eyepieces used.

Eyepieces

Eyepieces are briefly featured later – better eyepieces can give the binoculars two great assets:
• In 'wide-field' binoculars, they provide a greater field of view – considerably increasing the area of sky seen at one time. A normal pair of 8x40 binoculars might have a field of view of 6°, while that in a wide-field pair may be 8 to 9°.
• They can provide an increase in what is called the 'eye relief'. This determines how close the

Above right *The entire Southern sky, with the centre of the Milky Way passing near the zenith, is captured by a 180° fish-eye lens, from Lake Titicaca, Bolivia (South America).*

eye has to be to the back of the eyepiece to see the full field of view. If the eye relief is small, spectacle wearers will only view part of the field. Binoculars with long eye relief (around 20mm) allow the whole field to be viewed while wearing spectacles.

Image-stabilized binoculars

Owing to greater aperture and higher magnification, 10x50 binoculars can be excellent for astronomical use as they will help you to see fainter stars than 8x40, but they tend to be heavy and difficult to hold steady. Mounting brackets are available, enabling them to be used with tripods.

A new, but expensive, option is that of image-stabilized binoculars. These use a moving prism system controlled by a gyroscope to compensate for the small movements of your hands. You simply centre the object – say the Moon – in the field of view, press a button, and the image becomes totally steady. These binoculars allow higher powers to be used: 50mm lenses can be used with 15x and 18x powers, and so enable you to see fainter objects – but they do have a smaller field of view than lower power binoculars.

Telescopes

The objective

Every telescope will use either a lens or mirror to form an image. This is called the objective. The objective has a focal length which usually ranges from about 600mm to roughly 2000mm (23½–79in) in amateur telescopes. The image is then observed using an eyepiece which has a much shorter focal length – from about 3–56mm (up to 2in). The diameter of the objective is used to specify the size of a telescope – e.g. a 3in or 8in telescope has a mirror or lens of that diameter.

Though focal lengths of objectives and eyepieces are virtually always quoted in millimetres, apertures are often quoted in inches.

Magnification

Magnification is determined by the ratio of the focal length of the objective (lens or mirror) to that of the eyepiece. So, if you use a 25mm (1in) focal-length eyepiece with an objective of focal length 1000mm (40in), you will get a magnification of 40.

One advantage of higher magnification is that it enables you to see the fainter stars. As stars are point sources (all the light is concentrated in a single point), the brightness of their images do not decrease with magnification, but that of the sky background does so making the stars stand out more easily. Magnification is easily over done. Increase the magnification gradually using different eyepieces until no more detail is seen. It is very rare that a magnification of over 200 is worthwhile.

Above *An illustration by C Graham that appeared in the* Chicago Tribune *in October 1893 portrays the Yerkes 40th refracting telescope – still the largest refractor in the world.*

Resolution

The fineness of detail that can be seen is called the resolution and is theoretically determined by the size of the telescope objective. In angular terms this is about 1 arc second for a 6in (150mm) telescope. It is inversely proportional to the telescope aperture, so a 12in (300mm) telescope would have a resolution of about 0.5 arc seconds.

Opposite *The refracting telescope uses a lens to bring light to a focus where the image is viewed through the eyepiece.*

Seeing

It is usually the condition of the atmosphere that limits image quality – typically to about 2–4 arc seconds; however, there can be brief moments when the atmosphere steadies, enabling finer detail to be seen. This effect of the atmosphere on image quality is called the 'seeing'. The seeing is usually rated on a scale of 1 (very poor) to 10 (perfect), devised and named after William H Pickering. With poor seeing, the highly magnified image of a star will appear to be broken up into many pieces dancing around in the field of view, while the image of the Moon will appear to shimmer.

Focal ratio

The focal ratio, or f-ratio, of a telescope is simply the effective focal length divided by the aperture. This varies from around f/4 to f/20.
- Short-focal-length telescopes allow wide fields to be observed but image quality tends to suffer, so these are best for wide field observations.
- Long-focal-length telescopes from about f/10 to f/20 give smaller fields of view and are perhaps

best suited for planetary and lunar observations.
- Those telescopes in the middle range, around f/7 to f/9, are good all-rounders. It is possible to buy 'focal reducers' which reduce the effective focal ratio of a telescope from say, f/10 to f/6.3, so extending the versatility of a high-focal-ratio instrument.

Optical quality

Image quality of a telescope is limited by the 'seeing' as mentioned before – which is why you shouldn't judge a new one on just one night's observing. On nights when the atmosphere is calm, the optical quality of the lens or mirror can determine what detail you might expect to see. Many advertisements will claim that a telescope is 'diffraction limited'. This means that the resolution is limited by the size of the aperture – this can be more or less the case, provided that the mirror or lens introduces errors of no more than one-quarter of the wavelength of visible light across its aperture. Under near-perfect atmospheric conditions, a lens or mirror made to greater accuracy (around one-sixth to one-tenth wavelength) will give slightly better images.

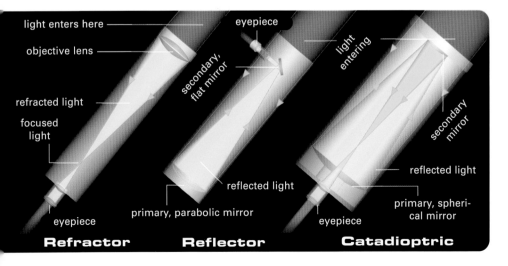

light enters here — objective lens — refracted light — focused light — eyepiece

Refractor

eyepiece — secondary, flat mirror — reflected light — primary, parabolic mirror

Reflector

light entering — secondary mirror — reflected light — primary, spherical mirror — eyepiece

Catadioptric

Brightness

Using a telescope will allow you to see fainter stars and other objects. The fainter stars will have a greater magnitude than the brighter (*see* p28), so a telescope will increase the magnitude that can be observed. This is called magnitude gain. The table below lists the magnitude gain that a telescope of a specific aperture will give. If, in a particular direction of the sky, you are able to observe a 4th-magnitude star with your unaided eye, then a 4in (102mm) telescope will allow you to observe an object 5.2 magnitudes fainter. Under ideal conditions you can see stars of magnitude 6.5 with your unaided eye, so by adding 6.5 to the figures in this table, you arrive at the limiting magnitude – that is, the faintest stars that will be visible through these telescopes. The limiting magnitude of the 4in telescope is thus around 11.7 (5.2 + 6.5).

TELESCOPE APERTURE		MAGNITUDE GAIN
3in	(75mm)	4.5
3.5in	(90mm)	4.9
4in	(102mm)	5.2
4.5in	(114mm)	5.9
5in	(130mm)	6.3
6in	(150mm)	6.7
7in	(180mm)	7.1
8in	(200mm)	7.4

Types of telescopes

Refracting Telescopes

Refracting telescopes use lenses to collect and bend light to the focal point. The first telescopes used a simple single lens to form the image. Simple lenses suffer badly from chromatic aberration – splitting the light up into its constituent colours – so producing coloured fringes around bright objects. A doublet lens made by combining one element of crown glass with one of flint glass was developed, called an achromat. These lenses very largely reduce false colour and are used in virtually all refracting telescopes today.

Achromatic refractors use an objective lens that attempts to bring light of all colours to a common focus but cannot achieve this entirely, resulting in some colour-fringing on the brighter stars and planets. The false colour largely takes the form of a purple halo around an object or perhaps a lime-coloured rim just inside the limb of the Moon.

The longer-focal-length (around f/8) versions are probably a better buy as these suffer less false colour. The 5in (130mm) aperture is a very good compromise between portability and light grasp (see Terminology opposite).

Apochromatic refractors use multi-element (two, three or four) lenses which virtually supress all false colour (the term apochromatic means 'free from spherical and chromatic aberration'). They tend to be very expensive but give exquisite images and will often outperform a somewhat larger reflecting telescope.

Opposite top *Examples of a reflecting telescope (top left) and a refracting telescope.*

Reflecting telescopes

Isaac Newton did not believe that chromatic aberration could be overcome, and therefore designed a reflecting telescope that used a mirror to focus the light; since the light doesn't pass through glass, no chromatic aberration is caused by the objective.

The Newtonian type is the simplest of all reflecting telescopes, so called after Isaac Newton's original design. Light is reflected from a parabolic primary mirror at the base of the telescope tube up to a flat secondary mirror close to the top end, which reflects the light through 90° to reach the focal plane, where the eyepiece is located. The performance of Newtonians is possibly the best of all types of reflectors. The focal ratios can vary from f/5.5 up to (rarely) f/10.

The smaller the focal ratio the wider the field of view; but very short-focal-ratio paraboloids suffer from a lens defect known as 'coma' – stars near the edge of the field of view look like little comets. A Paracorr device supplied by TeleVue can, however, provide excellent correction. Larger focal ratios provide a smaller field of view, but tend to provide higher image quality. A focal ratio of 8 is, as with almost all telescopes, an extremely good compromise.

TERMINOLOGY

Light grasp: Refers to the amount of light a telescope can collect – bigger aperture telescopes have a greater light grasp and can be used to observe fainter objects.

> **Parabolic mirror:** To bring the light rays from a star or planet to a sharp focus, the mirror has to have a specially shaped surface whose profile follows the mathematical curve of a parabola.
>
> **Parabola:** A mathematical curve with the property that if a mirror follows its outline, all the light rays arriving parallel to its optical axis are reflected to one point – its focus.

Catadioptric reflecting telescopes

These, somewhat more complex – and hence expensive – telescopes use a combination of a mirror and a lens (which is usually mounted at the front of the optical tube assembly) to form the image. The lens protects the mirror from dust or tarnishing and they can be very compact for their size so are understandably very popular.

A Schmidt-Cassegrain uses a secondary mirror near the top of the telescope tube to reflect light down through a central hole in the primary mirror at the base of the telescope – the Cassegrain focus. The Schmidt-Cassegrain is therefore compact because the light path is folded back on itself within the tube. The image is formed behind the primary mirror where the eyepiece is placed. A correcting lens – the corrector plate – mounted at the top end of the tube corrects for the fact that a spherical, rather than a parabolic, mirror is used. The disadvantages of this type of telescope are that the secondary mirror is relatively large, somewhat affecting image quality and a dew shield, warmed by electrical heaters, is often needed to prevent dew forming on the surface of the corrector plate.

A Maksutov-Cassegrain, like Schmidt-Cassegrains, uses a corrector plate at the front of the compact optical tube. The classic Gregory-Maksutov has its secondary reflector deposited on the rear face of a steeply curved corrector plate. Its performance can approach that of apochromatic refractors. Maksutovs, along with all other Cassegrain telescopes, tend to have relatively long focal lengths, limiting their field of view. They are superb for lunar and planetary observing, but not quite so adept at viewing the larger open clusters or sweeping the Milky Way.

Collimation

A telescope will only produce a good image on or close to its optical axis, so it is vital that this is in the centre of the field of view. The exercise to bring this about is called collimation.

Here is an outline of the procedure for reflecting telescopes:

- Point the telescope towards a white wall, remove any eyepiece and look through the focuser with your eye as close to its centre as possible.

- A 'sighting tube' made from a black film canister and a hole drilled through the centre of its base will fit in place of the eyepiece to enable you to centre your eye more precisely – putting crosswires across the far end of the canister can help as well.

- You will see in the secondary mirror a view of the primary mirror and its surroundings at the base of the telescope tube. The view should be totally symmetrical with the centre of the mirror in the centre of the field of view.

- If it is not, adjust the three screws that hold the secondary mirror in position until the view is symmetrical. Some mirrors have a black spot at the exact centre of their mirrors to aid this – you might like to put one on.

- Having adjusted the secondary mirror you should see your eye 'looking back at you'. If this is not exactly in the centre of the field of view then you need to adjust the three screws that determine the primary mirror axis until you see your eye precisely in the centre of the field of view.

Note: For an f8 Newtonian the above procedure should suffice but the correct collimation of Catadioptric telescopes is more critical and special accessories such as the 'Cheshire eyepiece' and laser based collimation tools can be bought to assist in the process.

Eyepieces

The eyepiece of a telescope (or binoculars) acts as a high-quality magnifying glass. The shorter the focal length, the closer your eye gets to view an image. This has two effects:

- the 'apparent' size of a viewed object (or the distance between two stars) is increased, so also increasing the magnification.

- the visible area of the telescope's image in the focal plane is reduced, thus reducing the actual field of view.

Apparent field of view

If you look through an eyepiece at a white surface, you will see a circular white field delimiting some angle. This is the apparent field of view normally measured in degrees. Today, most eyepieces give an apparent field of at least 50°.

Actual field of view

The 'actual' field of view is the angular size (in degrees) of sky you can see through a particular eyepiece. This can be calculated approximately by dividing the apparent field of view by the

Above *A range of eyepiece types of roughly the same focal length (from left to right): 9.7mm Super Plössl, 9mm Nagler wide-field, 9.5mm Plössl, and 9mm Kellner.*
Opposite top *A Maksutov-Cassegrain telescope.*

magnification. (So with a 50° field of view eyepiece and a magnification of 100 one would see a half-degree wide actual field of view.)

Wide and super wide-field eyepieces

Eyepieces giving wider apparent fields of view, i.e. greater than 50°, are available, but these are expensive! So-called 'wide-field eyepieces' give apparent fields of around 65°, super wide-field eyepieces give fields of view of around 80° or more. One advantage of very wide-field eyepieces is that they allow you to observe, say, a view encompassing a globular cluster at a higher magnification than with a narrower-field eyepiece; the higher magnification reduces the apparent brightness of the sky background, so enabling fainter stars to be seen.

Above *Wide-angle designs give apparent fields of view of around 65°, which show 70% more sky than Plössl of the same focal length.*

Eye relief

This defines the distance at which the pupil of the eye has to be placed behind the eyepiece to gather all the light passing through it. A rubber extension to the eyepiece usually indicates this best position for normal eyesight. Generally, the shorter the focal length of the eyepiece, the less the eye relief. Some people find it hard to use eyepieces with very small eye relief, and it certainly makes it impossible to wear glasses. For those, perhaps with severe astigmatism, who need to wear glasses when using a telescope, longer focal-length eyepieces can be used in conjunction with a Barlow lens (which usually doubles the effective magnification) or alternatively, eyepieces

Above *Vixen Lanthanum has a range of eyepieces of focal lengths ranging from 2.5–25mm; their unique feature is an eye relief of 20mm.*

The Barlow lens typically doubles the magnification of an eyepiece.

Above *Eyepieces with barrel diameters of up to 50mm (2in) enable a greater area of the image to be viewed.*

Above *This 6x30 achromatic finder scope from Orion makes use of a four-element Plössl eyepiece.*

such as the Vixen Lanthanum range can be used which are designed to have a fixed 20mm (0.8in) eye relief at all focal lengths.

Filters

Some objects, such as supernovae remnants and planetary nebulae, give off a lot of their light at specific wavelengths (spectral lines) such as those of excited hydrogen and oxygen. One can buy filters, particularly those termed UHC (for Ultra High Contrast) or OIII (to pick out the excited Oxygen lines) that allow light at these wavelengths to pass but reduce all other light including light pollution from sodium lamps. They thus greatly increase the visibility of these objects. Unless one has very dark skies their use may be the only way that, for example, the Veil Nebula can be seen.

Finder scopes

As telescopes have quite a small field of view (around 1°), they are normally equipped with a 'finder scope' mounted on the main telescope. This has a wider field of view (around 5°) and enables you to align the telescope on objects sufficiently accurately that they may also be seen through the main telescope eyepiece. Finder scopes have always been small telescopes similar to one-half of a pair of binoculars, and using the same nomenclature (say, 9x50) but with an inverted image. The eyepiece of the finder includes an etched glass or wire cross to indicate the centre of the field. A second type of finder is becoming quite common: when you look through it, a red dot appears to hang in the sky and you simply move the telescope until the dot lies over the desired target object. This finder type is great for planets and the brighter stars, but it actually reduces the brightness of objects so cannot be used for seeking out faint galaxies.

In both cases it is necessary to properly align the finder before you use it. Locate an object either in daylight, for example the top of a distant tree, or at night – the Moon or a bright planet. Find the object with the main telescope, then lock the tube's position. Look through the finder and adjust its positioning screws to centre the object on the crosswires.

There are two types of telescope mount used with amateur telescopes: the Altazimuth and the Equatorial mounts.

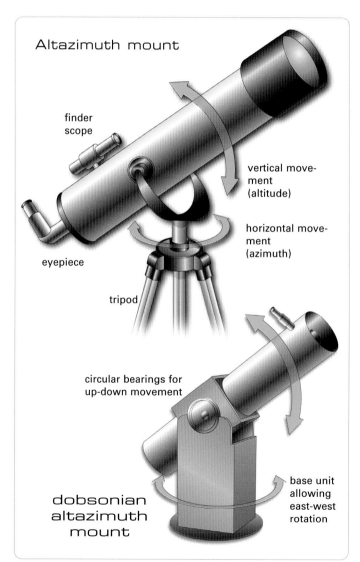

Altazimuth mount

Altazimuth mount

finder scope

vertical movement (altitude)

horizontal movement (azimuth)

eyepiece

tripod

circular bearings for up-down movement

base unit allowing east-west rotation

dobsonian altazimuth mount

Altazimuth mount

This is a simple mount supported on a tripod, generally used for refractors up to about 4in (100mm) in aperture (altitude = up or down; azimuth = around the horizon). They work well and high-quality ones are used by owners of very expensive apochromatic refractors.

Another type of altazimuth mount is the Dobsonian mount, primarily used for Newtonian reflectors. The telescope is simply pushed to follow an object in an amazingly smooth motion. As the mirror is so close to the ground, this mounting is very stable. Altazimuth mounts are not suitable for long-exposure astrophotography but you can take short-exposure images of the Moon.

Equatorial mount

Equatorial mounts allow an object to be tracked across the sky using a motor drive system; they are thus suitable for astrophotography. This type of mount has a polar axis arm that is adjusted to point to the north (or south) celestial pole, with a declination axis mounted on it. This is rotated to set the declination of the object you wish to observe. Simply move the telescope to point at the object using the finder, and then lock the declination clamp.

By rotating the polar axis at the sidereal rate (the apparent movement of the stars) in the appropriate direction (opposite in each hemisphere), the object will stay in the field of view if the mount has been aligned. You can do this manually, or have an RA (Right Ascension) motor drive fitted, which automatically turns the polar axis at the sidereal rate. (If the polar axis is slightly misaligned, small corrections to bring the object back in the field of view can be made as required with fine RA and Dec controls.)

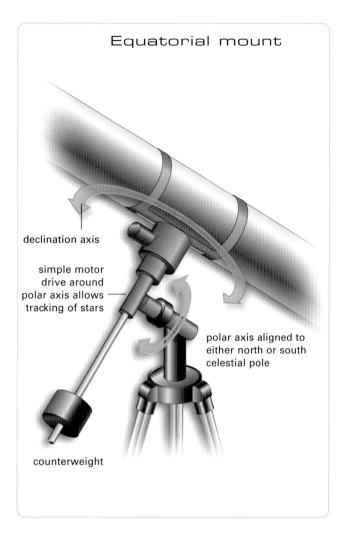

Equatorial mount

declination axis

simple motor drive around polar axis allows tracking of stars

polar axis aligned to either north or south celestial pole

counterweight

Fork mount

Almost all Schmidt-Cassegrain and Maksutov telescopes are mounted on the supporting arms of a fork mount, directly attached to the tripod; in this case they have to drive in both azimuth and altitude to track an object, so this is effectively another form of altazimuth mount. (Many such altazimuth mounts are now used in the so-called 'go-to' telescope mounts.) Because of the two fork arms, no counterbalance is required. The fork supporting the telescope can also be mounted on an equatorial wedge above the tripod. This effectively turns it into an equatorial mount; in this case, only an RA drive is required to track an object.

Go-to mounts

Computer-controlled telescopes have become very popular. These telescopes are usually equipped with a fork mount in altazimuth mode. The telescope's position on the Earth's surface (latitude and longitude) and the time are set into the linked hand-held computer. This information can be obtained highly accurately with a GPS receiver used for navigation, so an increasing number of telescopes include one to remove even this task. The mount must then be levelled and pointed north to reasonable accuracy (some mounts even include an electronic compass!), and the setting-up procedure begins.

The computer decides on two or three bright stars, well spaced around the sky, that will be above the horizon at the time of observation; it directs the telescope to the first star (probably not too accurately unless the telescope setup was very carefully done), then you manually centre the star in the field of view and confirm when you have done this. The telescope then slews to the second star for manual centering and confirmation. The computer can then calculate any errors in pointing the telescope due north, and tilt in the telescope base, so is then able to slew the telescope to any object in its extensive database. This will certainly be within the field of a low-power eyepiece.

You can even ask the computer to take you for a celestial tour and it will seek out the most interesting objects in the sky at that time!

Fork mount

fork arms support telescope

motor drive allows tracking of stars in altitude

equatorial wedge attached to tripod aligns fork axis toward relevant pole

Above *This 8in (200mm) Newtonian telescope mounted on a Dobsonian Altazimuth Mount is pushed by hand to track an object across the heavens.*

Telescope maintenance

Telescopes do not wear out, but they do need looking after!

- When in use, try to keep the lens or corrector plate, if any, dry by using a dew shield. Bringing a telescope into a warm room from the colder outdoors can cause a problem as dew may form on the glass surfaces. To avoid this, you can cover the telescope optics outside, then wrap up the telescope to slow down any heat transfer and bring it into as cold a room as possible. If it does dew up, keep it uncovered until it fully dries.
- Store the telescope in a dry but preferably cool place; the latter will reduce the time taken for the telescope to cool down to the outside temperature when used.
- It obviously makes sense to protect the mirror or lens from dust by ensuring that the telescope is stored with a cover over the mirror or tube assembly.
- Over time the mirror of a Newtonian, and the objective or corrector plate of a refractor or catadioptric telescope, will get dirty and the surface will need to be cleaned.

First blow off all the dust you can with an air pump like those used for pumping up airbeds. Then wash them very gently using cotton wool (sterilized surgical cotton is best) and distilled water or a mix of 75% distilled water and 25% isopropyl alcohol. Hold the lens or mirror vertical so that the cleaning water runs off, then dab off any that remains with cotton wool. Owners of Newtonian telescopes should expect that after a few years of use there may come a time when the mirror needs to be re-aluminized.

Your own observatory

Telescopes with apertures of up to about 11in (279mm) can usually be handled by one person, so these can be stored indoors and taken out into the garden when required. What can deter people using them, however, is that time has to be spent setting up and aligning the telescope. Its temperature then needs to adjust to the outdoor temperature before good images are obtained as, if the telescope and the air within it are not at the same temperature as the surroundings, there will be air currents moving around that can totally destroy the optical images. It is thus important to allow your telescope to come into equilibrium with its environment and this can take two hours or more if taken from a warm room out into a cold atmosphere. Refractors tend to settle down more quickly than more complex types. Obviously, observing would be far easier if the telescope were constantly ready for viewing, but this requires building an observatory!

It is possible to buy observatory domes made from fibreglass. These can be erected over a rotating base; they have a door that slides up and back to reveal an opening for the telescope. At lower cost, one can use variants on a garden shed. The entire shed – with a suitable large door at one end – mounted on rails, can then be moved away from the telescope. Alternatively, the shed can be fixed but its roof moves sideways along two runners that extend to one side of the shed. Another alternative is a 'glasshouse' type of construction on a similar rotating base, where the door hinges open to one side.

In all cases, you may be plagued by street and security lights. It's possible to make, or buy, black screens that are placed between your telescope and the offending light source or even a black bag that can be placed over the offending light temporarily. This will aid the dark adaptation of your vision and can be very worthwhile.

Right *Three versions of a home observatory. Some amateur astronomers build their own observatory, but it is also possible to obtain these from specialist companies.*

wooden shed on rails

split wooden shed on rails

door flap slides back

opening

telescope

fibreglass dome

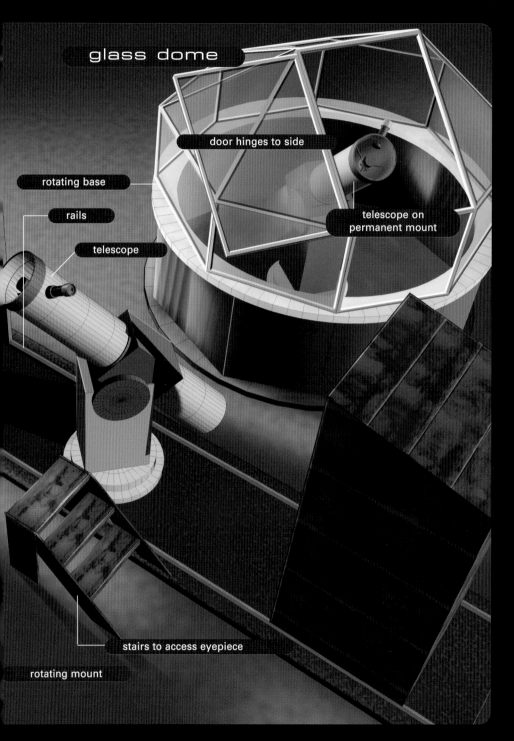

glass dome

door hinges to side

rotating base

rails

telescope

telescope on
permanent mount

stairs to access eyepiece

rotating mount

4

OBSERVING THE SOLAR SYSTEM

The planets of the solar system were formed at the same time as the Sun about 4.5 billion years ago. As a giant cloud of dust and gas contracted under the influence of gravity, the central concentration, which formed the Sun, became surrounded by a flat disk of material rotating around it. Gradually the heavier particles began to form larger clumps which eventually become what are called 'planetismals'. Collisions and gravitational attraction made them grow, eventually forming the planets. The Sun's radiation pressure swept most of the lighter gases out from the inner solar system leaving only heavier particles that eventually formed the terrestrial planets. Further out, planets still first formed rocky cores, but were able to attract and hold onto the helium and hydrogen gas in their vicinity so they became much bigger, so giving the gas giants.

Opposite *This illustration (not to scale) outlines, in the background, the orbital routes of the Sun and nine planets within the solar system.*

The Moon

The Moon and its phases

The Moon is the Earth's only natural satellite and, as with all objects in the solar system, (apart from the Sun itself) it gives off no light of its own. Every month the Moon orbits the Earth, completing a cycle of lunar phases from one New Moon to the next, a period known as the synodic month. The New Moon is the time when the Moon is directly between the Earth and the Sun and is not visible to us and, although you cannot see it, the New Moon is in the sky all day. We commonly refer to the first visible crescent of the Moon seen just after sunset as the New Moon and are then often able to see the unlit part of the Moon because the lunar surface is faintly illuminated by sunlight reflected onto the Moon from the Earth, known as 'earthshine'. The Moon moves about 13° eastward each day in its monthly orbit, and is seen first as a thin crescent close to the point where the Sun has set, then the visible area grows via a gibbous phase to a Full Moon. As the days progress, the Moon is visible in the sky for longer and more of the disc is illuminated. At First Quarter, following a New Moon, half of the Moon is visible. At Full Moon, the Moon rises at sunset, and sets at sunrise. The phases then occur in reverse and at Last Quarter, one-half of the face is visible again.

Observing the Moon

Because the Moon rotates once on its axis in the time it takes to orbit around the Earth, we always see the same side of the Moon – its nearside has become 'locked' towards the Earth. However, as its orbit around the Earth is not a perfect circle, it appears to wobble slightly back and forth, allowing us to see a little around the sides, an effect known as libration. In time, this enables us to see nearly 60% of the Moon's surface. Regular observations of features, such as Mare Crisium, show how they are

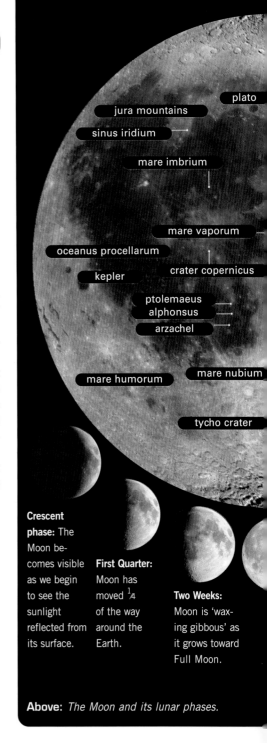

plato

jura mountains

sinus iridium

mare imbrium

mare vaporum

oceanus procellarum

crater copernicus

kepler

ptolemaeus
alphonsus

arzachel

mare nubium

mare humorum

tycho crater

Crescent phase: The Moon becomes visible as we begin to see the sunlight reflected from its surface.

First Quarter: Moon has moved $\frac{1}{4}$ of the way around the Earth.

Two Weeks: Moon is 'waxing gibbous' as it grows toward Full Moon.

Above: *The Moon and its lunar phases.*

THE SYNODIC MONTH

The Moon's orbit around the Earth in 27.3 days is termed its sidereal period; i.e. its motion relative to the background stars. Owing to the Earth and Moon's motion around the Sun, it takes slightly longer for the Moon to present the same phase to us; i.e. from Full Moon to Full Moon. This period is the synodic lunar month of 29.53 days and is what we refer to as a lunar month.

mare crisium

mare serenitatis

mare tranquillitatis

mare fecunditatis

mare nectaris

Last Crescent: the side of the Moon that faces Earth is invisible to us

Fourth Week: Moon is at Last Quarter. and visible in the morning sky.

Third Week: Moon is 'waning gibbous' in the days following Full Moon.

Full Moon: rises 12 hours after the Sun & is directly opposite the Sun and visible throughout the night.

sometimes closer to the limb than at other times. The features on the Moon's surface are best seen when they are near the terminator – the edge of the illuminated part of the surface – as here shadows cast by the Sun are longest. At Full Moon, when there are no shadows, features are not so easily seen, but the 'rays' of ejecta thrown out when a crater is formed show up well; those of the young crator Tycho are particularly prominent.

Surface features

The lunar surface is of two distinct types: the relatively bright and heavily cratered highland regions and the darker, less cratered maria (Latin for 'seas'). The maria were formed by giant impacts which hollowed out the surface and allowed darker, basalt-type rocks to well up from under the crust. Because they were formed near the end of the period when debris from the formation of the solar system was impacting onto the planetary and satellite surfaces, they show less cratering. The maria have been given evocative names like Mare Imbrium, Mare Tranquillitatis and Mare Fecunditatis.

Lunar eclipses

Once every 14 months, on average, the Full Moon falls into the Earth's shadow and we have a total eclipse. You might expect that the Moon would totally disappear as it only shines by

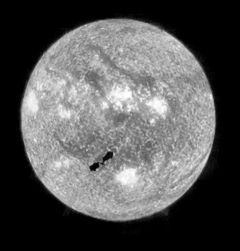

reflected sunlight, but we do see the Moon in eclipse because there is sufficient light scattered through the Earth's atmosphere to dimly illuminate it. What we will see during the eclipse is very dependent on the amount of dust in the atmosphere. If the atmosphere is not too dusty, then we see a beautiful russet-brown coloration but if, perhaps after a volcanic eruption, the atmosphere is laden with dust, the Moon will appear darker and have a rather dull grey colour. The next total eclipses of the Moon will occur on the 3 March 2007, 28 August 2007 and 21 February 2008.

The Sun

By far the largest object in the solar system, the Sun makes up more than 99.8% of its total mass. It is our own star and, like all other stars, gets its energy from nuclear fusion which is produced in the core – here the temperature is 15.6 million °C (28 million °F). The energy from the core reaches the surface called the photosphere where it is emitted into space. Above the photosphere lies a thin zone called the chromosphere beyond which is the extended outermost part of the Sun called the corona. The light from these zones is very faint in comparison to that from the photosphere, they are only visible through special telescopes or during solar eclipses.

Sunspots and the rotating Sun

When the Sun's magnetic field breaks through the photosphere; it prevents the convection of hot gas from deep within the Sun, creating regions of lower temperature. Because these regions are a little

Opposite bottom *Sunspots are proof of magnetic activity at deeper levels within the Sun. They often appear in groups, changing in shape and size as they cross the solar disc.*

SUN FACT FILE

❋ **Distance from the Earth:**
Approx. 150 million km
(93 million miles)

❋ **Size:**
Diameter 1.4 million km
(870,000 miles)

❋ **Polar (75˚ latitude) rotation period:**
33.4 Earth days

❋ **Equatorial rotation period:**
25.7 Earth days

❋ **Temperature of corona:**
Rises to 2 million °C (3.6 million °F)

❋ **Surface temperature:**
470°C (870°F)

❋ **No. of planets in solar orbit:**
9

cooler, they appear dark in comparison with their surroundings and are thus called 'sunspots'. The number of sunspots fluctuates in a cycle that lasts for approximately 11 years. During a period of about seven years the number of visible sunspots slowly increases up to what is called solar maximum. The field then begins to break down and the number of sunspots reduces toward solar minimum when the whole eleven-year cycle repeats – though with the magnetic field reversed. This is called the solar cycle. Near solar maximum the Sun may produce many solar flares, streams of charged particles that are ejected into space. When these interact with the Earth's atmosphere we see beautiful displays of the Aurora – bands of white, green and red light arching across the sky.

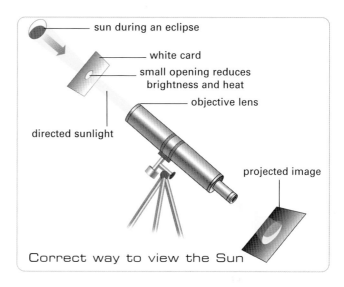

sun during an eclipse

white card

small opening reduces brightness and heat

objective lens

directed sunlight

projected image

Correct way to view the Sun

Observing the Sun

NEVER look directly at the Sun! Rather direct the Sun's light into the objective lens of a telescope, and then project its image onto a sheet of white cardboard held a few inches away from the eyepiece (see illustration). To reduce the Sun's heat radiation which could damage the eyepiece, 'stop down' the aperture to no more than 25mm (1in) or so across – many telescopes have covers with just such an aperture for this use. Use the shadow of the telescope or finder to line up the telescope toward the Sun. DO NOT look through the finder or along the telescope axis! Then, by adjusting the focus of the telescope, you will be able to obtain an image of the Sun a few centimetres across. If you observe sunspots on the disc and watch them over a number of successive days you will find that they move across the disk, growing and shrinking in size and taking 25.7 days to make one complete rotation at the equator.

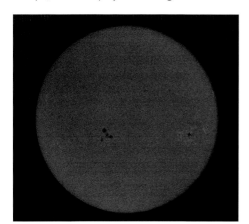

Solar eclipses

They occur at New Moon when the Moon happens to exactly line up with the Sun and its shadow is cast upon the Earth's surface blocking out the light from the Sun. Just as with lunar

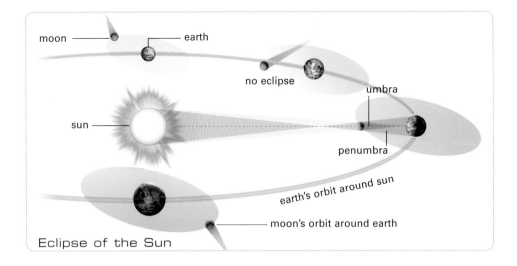

Eclipse of the Sun

eclipses, the inclination of the Moon's orbit means that they do not occur every month, but there is one visible somewhere on the Earth's surface about every two years.

The Moon's orbit is elliptical and so its distance from the Earth varies; the nearer the Moon, the larger the extent of its shadow on the Earth and the longer the observed eclipse. Points on the equator are nearest the Moon and so eclipses seen here will have the maximum possible period of totality – over seven minutes when the Moon is at its closest to us in its orbit. On the other hand, if the Moon is at, or nearing, its greatest distance from us, it does not have a sufficient angular size to totally cover the Sun and we see a bright ring surrounding the dark disc of the Moon – called an annular eclipse.

As the Moon eclipses the Sun, you will see the chromosphere – a lovely red-pink colour due to the glowing hydrogen gas within it. Often prominences – material carried upward by the Sun's magnetic field – may be seen arcing out into space. Surrounding this will be the pearly-white glow of the Sun's corona – a most beautiful sight.

Mercury

Mercury is known as an inferior planet because it is closer to the Sun than the Earth. It orbits the Sun in 88 Earth days; this is the Mercurian year. Its orbit is elliptical so the distance between the planet and the Sun varies from 0.31 to 0.47 AU. In comparison to the Earth, Mercury spins on its own axis very slowly, taking 58.6 Earth days to complete one rotation relative to distant stars. So Mercury spins on its axis three times in the time it takes to orbit the Sun twice. To anyone standing on the surface of Mercury, the Sun would rise in the sky only once every 176 days (after two orbits). The temperature on this planet can vary from 470°C (800°F) to around –180°C (-300°F). Mercury's atmosphere is very thin – consisting of negligible amounts of oxygen and sodium, with traces of helium, potassium and hydrogen captured from the solar wind.

Surface features

Mercury never gets more than 20° or so from the Sun, so the planet is often lost in the twilight, making it difficult to observe the planet's features. Most of our knowledge of this planet comes from the close approach of NASA's spacecraft Mariner 10 which flew by Mercury three times in 1974 and 1975. The surface features show that Mercury resembles our Moon in appearance, with many craters and lava plains called maria. The planet also has long sinuous features, called 'lobate scarps', which are between 20 and 500km (12 and 300 miles) long and rise to a height up to 2km (1 mile). These are wrinkles in the surface – thrust faults – created as the planet cooled and shrank. Surprisingly, radar observations have shown that, in craters close to the poles where the Sun's heat never reaches, lie deposits of ice, presumably the result of the impact of comets.

Observing Mercury

Binoculars will be a useful aid to see Mercury in the sky – it can be distinguished from any bright stars that might be in the vicinity by the fact that it does not scintillate, or twinkle, to the same extent. When the prospects are good, one needs to find a location with a very low western (if sunset) or eastern (if dawn) horizon. If Mercury is to be observed in the evening, preferably get to your location before the Sun sets as this will help you to know where to look for Mercury – a little up from where the Sun has disappeared. Sometimes Venus or the thin crescent Moon may be close, helping to locate it. Monthly sky guides will alert you to good opportunities to observe it. For observers away from the Equator, it is best seen in the spring or autumn when the angle of the ecliptic to the horizon is greatest and the Sun sets (or rises) most steeply toward the horizon. Then, if Mercury is close to maximum elongation – that is, furthest in angle from the Sun – it will have a reasonable elevation at sunset or before dawn.

MERCURY FACT FILE

✷ Relation to the Sun:
Closest to Sun
Mean distance 57.9 million km
(36 million miles)

✷ Relative size:
Second smallest of planets

✷ Atmospheric composition:
Negligible atmosphere

✷ Orbit around the Sun:
88 Earth days

✷ Period of rotation:
58.6 Earth days

✷ Surface temperature:
470°C (870°F)

✷ No. of moons:
0 moons

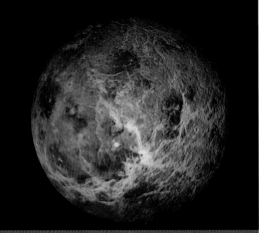

VENUS FACT FILE

❖ **Relation to the Sun:**
Second planet from the Sun
Mean distance 108 million km
(67 million miles)

❖ **Relative size:**
Very similar to Earth

❖ **Atmospheric composition:**
96% carbon dioxide, 3% nitrogen,
traces argon, water

❖ **Orbit around the Sun:**
224.7 Earth days

❖ **Period of rotation:**
243 Earth days

❖ **Surface temperature:**
480°C (896°F)

❖ **No. of moons:**
0 moons

Venus

Like Mercury, Venus is also an inferior planet lying between the Earth and the Sun. It is the brightest planet, and the third brightest object in the sky after the Sun and Moon. It is sometimes called the 'morning' or 'evening' star as it is seen in the hours before dawn or after sunset. It rotates on its axis in 243 Earth days making the length of its day relative to the stars longer than its year (224.7 Earth days). Added to this, it rotates in a retrograde sense (i.e. in an opposite direction to its movement around the Sun). These two motions combine to give a Venusian day of 117 Earth days.

Above *A computer generated image from data taken by the Magellan spacecraft of Sapas Mons (foreground) and Maat Mons (background).*

Venus's Greenhouse effect

The surface of Venus is never seen from Earth because it is always covered by a thick, dense layer of cloud, believed to consist of sulphuric acid and particles of sulphur, probably originating in the surface volcanoes. Because of the abundance of carbon dioxide in the atmosphere, Venus suffers from the greenhouse effect – while incoming light and heat from the Sun pass through the atmosphere, longer wavelength infrared radiation cannot escape – resulting in a surface temperature of 480°C (896°F). The planet is very dry with little water. The slow rotation, thick atmosphere and high-velocity winds even out the day-time and night-time temperatures.

Surface features

Radar studies from Earth and the Magellan spacecraft in orbit around Venus have shown that most of Venus consists of gentle rolling plains, some surface depressions and two large highland areas. The highland areas called Ishtar Terra, in its northern hemisphere and Aphrodite Terra straddling the equator, are both the size of continents on Earth. The highest mountain on Venus, called Maxwell Montes, rises about 11km (7

Right *The circular corona appears in the southern Aphrodite Terra. Magma welled up, then retreated, after which the surface collapsed to form concentric lines of ridges and faults.*

miles) above the mean surface level at the eastern end of Ishtar Terra. Volcanoes dominate the surface; many show signs of recent activity and much of the planet is covered in lava flows.

There are some large craters on Venus – but no small ones, as small meteoroids burn up in the thick atmosphere before they can reach the ground.

Observing Venus

Venus can hardly be missed as it dominates the sunset or dawn sky. Extending up to 47° away from the Sun, it can be seen for three or four hours before sunrise or after sunset. One very interesting observational fact is that, even though Venus shows phases just like the Moon, the brightness hardly varies. It just so happens that, as the crescent narrows when Venus comes closer to the Earth (which would reduce the brightness), the increase in angular size almost exactly compensates, so giving a magnitude close to -4 for much of the time. In observing the changes in phase and angular size today, you are repeating the observations made by Galileo centuries ago that proved the Sun was at the centre of the solar system.

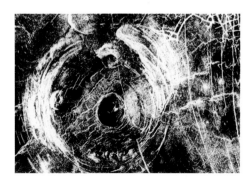

71

MARS FACT FILE

❋ **Relation to the Sun:**
Fourth planet from the Sun
Closest planet to the Earth
Mean distance 228 million km
(142 million miles)

❋ **Relative size:**
Half the size of Earth

❋ **Atmospheric composition:**
95% carbon dioxide

❋ **Orbit around the Sun:**
687 Earth days

❋ **Period of rotation:**
24 hours 37 minutes

❋ **Surface temperature:**
-65°C (-85°F)

❋ **No. of moons:**
2 moons

Mars

Being roughly half the size of Earth, Mars' interior has cooled more quickly and there is no current volcanic activity. Mars, popularly known as the Red Planet as a result of iron oxides on its surface, has an iron-rich core measuring 2900km (1800 miles) across, a mantle 3500km (2175 miles) thick and a crust 100km (60 miles) thick. Mars is similar to the Earth in that it rotates in 24 hours 37 minutes about an axis tilted at 24° to its orbital plane, so giving rise to similar seasons. The surface details, which include white icy polar caps, made up of frozen carbon dioxide and water ice, and dark features such as Syrtis Major change with the seasons as the ice caps expand and contract and as dust blows across the surface. The atmosphere is largely carbon dioxide and the atmospheric pressure is just one hundredth that of the Earth.

Observing Mars

Mars is a wonderful object for small telescopes, but only really worth observing when it is at opposition (opposite the Sun in the sky), at which time we are closest to the planet. This occurs once every two years or so. However, as Mars has an elliptical orbit, its distance to us at closest approach can vary widely, which means that the observed angular size, and hence visible detail, will vary too. In August 2003, Mars was at its closest to us for nearly 60,000 years, giving an angular size of 25 arc seconds – but it can be as little as 13.5 arc seconds. Something of Mars' surface is visible at such times. Against an overall salmon-pink coloration, the most obvious markings are the white polar caps and the dark triangle of Syrtis Major. At the closest oppositions more subtle markings can be seen, but the sur-

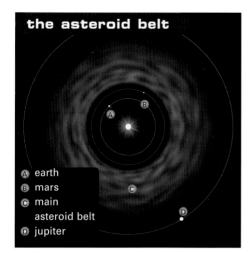

the asteroid belt

- ⒶＡ earth
- ⒷＢ mars
- ⒸＣ main asteroid belt
- ⒹＤ jupiter

The Gas Giants

Jupiter and Saturn are the largest planets in our solar system. They can be seen all year and as they are at great distances from the Sun, their apparent sizes do not alter so dramatically as we circle the Sun within their orbits. Their overall angular sizes at opposition, if the rings of Saturn are included, are very similar at 45 arc seconds, thus allowing a wealth of detail to be seen.

Jupiter

face detail can be totally obliterated by occasional giant dust storms lasting several weeks – as Mars' surface is obviously at its hottest when closest to the Sun. The darker markings are in the Southern Hemisphere, seen toward the top of the planet in a telescopic view. As the planet rotates in just over one Earth day, the view will change over a few hours.

THE ASTEROIDS

Between Mars and Jupiter lies the Asteroid Belt. Some of the asteroids can be seen with a telescope and one, Vesta, is visible to the unaided eye. They look starlike, however, and the precise location is needed within a detailed sky chart to be sure that you have actually seen one. One of the computer planetarium programs is almost a necessity; these will give the exact location and magnitude, enabling you to match the field of view with the computed map.

Right A montage of Saturn with its four brightest moons.

Jupiter has a creamy yellow coloration against which can be seen the darker equatorial bands which, if seeing conditions are good, may show whirls and spots, some lighter, some darker than the bands. The most famous is, of course, the Great Red Spot, a storm in the Jovian

⚛ **Relation to the Sun:**
5th planet from the Sun
Mean distance 778 million km
(484 million miles)

⚛ **Relative size:**
Largest planet in the solar system
Diameter is 11 times that of Earth

⚛ **Atmospheric composition:**
86% hydrogen, 13% helium,
clouds of methane and ammonia

⚛ **Orbit around the Sun:**
11.9 Earth years

⚛ **Period of rotation:**
10 hours

⚛ **Surface temperature:**
-110°C (-170°F)

⚛ **No. of moons:**
60 moons

atmosphere twice the width of our Earth! It does not always look so red and while sometimes it is very prominent, at othe times it is hardly visible – just a bite out of the band in which it lies. Jupiter is always changing, so one never tires of observing it. Any telescope will show the four Galilean satellites; in order of their distance from Jupiter, they are: Io, Europa, Ganymede and Callisto. The number that is visible changes every night as the satellites move behind or in front of the planet's disc. If seeing conditions are very good they can sometimes be just detected when lying in front of the disc, and often one of their shadows will fall onto the surface, appearing as a dark spot. Under good conditions one can even make out the disc shapes of the moons and it is even possible to observe the orange coloration of Io.

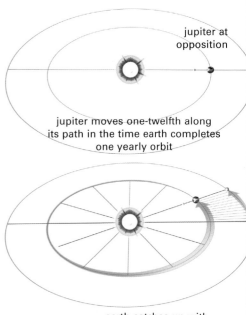

jupiter at opposition

jupiter moves one-twelfth along its path in the time earth completes one yearly orbit

earth catches up with jupiter after 13 months and is again at opposition

The four Galilean satellites, in order of their distance from Jupiter, are: Io (diam. 3630km; 2255 miles), Europa (3138km; 1950 miles), Ganymede (5262km; 3270 miles) and Callisto (4800km; 2980 miles), all named after lovers or attendants of Zeus, – and who is Jupiter's equivalent in Greek mythology. All except Europa are bigger than the Moon, and each is very different.

Io
Closest to Jupiter, Io (top right) experiences huge tidal forces as it orbits the planet, during which its surface moves in and out by about 100m (330ft). This generates a lot of heat, which is probably the cause of Io's numerous active volcanoes.

Ganymede
The solar system's largest satellite, Ganymede (not illustrated) is a mainly icy body; its mass is less than half of Mercury's. It has a small solid core, surrounded by a rocky silicate mantle and an icy surface.

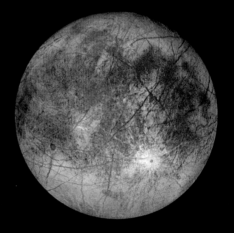

Europa
Europa (right) has an icy surface with little cratering but featuring giant fractures, and there may be lakes of liquid water beneath the surface. Europa is considered to be one of the more likely places for life existing elsewhere in the solar system.

Callisto
This is the only Galilean moon (right) not to show signs of geological activity on its surface and it has the most cratering. The craters are distinct from those seen on the Moon.

SATURN FACT FILE

❂ **Relation to the Sun:**
6th planet from the Sun
Mean distance 1427 million km
(887 million miles)

❂ **Relative size:**
Second largest in solar system

❂ **Atmospheric composition:**
96% hydrogen, 3% helium, traces of
methane and ammonia

❂ **Orbit around the Sun:**
29.5 years

❂ **Period of rotation:**
10 hours

❂ **Surface temperature:**
-140°C (-220°F)

❂ **No. of moons:**
30 moons

Saturn

Saturn is simply beautiful. There are faint bands
to be seen on the surface and sometimes white
spots occur. Saturn is far less dynamic than
Jupiter, but this is amply compensated for by the
wonderful ring system that surrounds it. A small
telescope will show three rings. Working inwards,
we have the outer ring (A), the bright ring (B) and
the inner ring (C), usually called the crepe ring.
This is fainter and a lot harder to see. Between
the A and B rings a dark gap is seen; this is
referred to as Cassini's division.

The ring system, made up of myriad particles of
ice and no more than 1km (1/2 mile) in thickness,
is tilted to the plane of the ecliptic. You view the
rings from above and below as the planet orbits
the Sun every 29.46 years. When seen edge-on,
they virtually disappear. The rings became fully
open in 2003, so will be good to observe for several years to come. Saturn has five satellites that
can be easily viewed with an amateur telescope.
The brightest, Titan, at magnitude 8.4, is easy to
spot, but the other four – Rhea, Tethys, Dione and
Enceladus – may be difficult to distinguish from
stars. The use of a computer program showing
stars down to magnitude 11 will be a great help

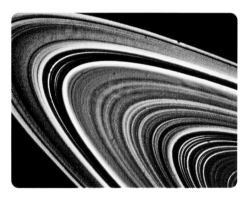

Above *A voyager image of Saturn's ring system
showing the intricate structure within them.
The dark band is Cassini's division.*

in identifying them. Monthly charts appear in astronomy magazines, showing their positions relative to Saturn.

Uranus

Uranus was the first planet to be discovered in modern times – by William Herschel in 1781 in the city of Bath where he was an organist. Most of our information about Uranus was obtained when Voyager 2 visited the planet in 1986; during its flyby, 10 of the planet's smaller moons were revealed. Uranus has 21 known moons and is also surrounded by 11 faint rings – the outermost ring is called the Epsilon ring. The planet has an angular size of 3.7 arc seconds and has a magnitude of 6. Uranus is a turquoise-green colour and, under perfect skies, you might even spot it with the unaided eye.

Geology of Uranus

Uranus's rocky core is surrounded by water, ammonia and methane ices, and an atmosphere consisting of mainly hydrogen. The planet's blueish appearance is caused by the atmospheric methane gases, which absorb red light thus giving off a green-blue reflection.

A toppled planet

Uranus's axis of rotation is tilted at 98°, making it lie almost on a level with its orbital plane around the Sun. It is believed that the planet may have collided with another body, toppling it onto its side. This leads to extreme seasonal changes as the north and south poles point directly at the Sun during the course of the planet's year – which effectively results in 42

URANUS FACT FILE

※ **Relation to the Sun:**
7th planet from the Sun
Mean distance 2870 million km
(1783 million miles)

※ **Relative size:**
About one-third the size of Jupiter

※ **Atmospheric composition:**
83% hydrogen, 15% helium,
2% methane

※ **Orbit around the Sun:**
84 years

※ **Period of rotation:**
17 hrs 14 mins

※ **Surface temperature:**
Av. -195°C (-320°F)

※ **No. of moons:**
21 (known)

years' daylight and 42 years' darkness. Since 1994, the Hubble Space Telescope has observed the planet in visible and infrared light, showing rapidly rotating zones and high-latitude clouds developing as the planet's northern hemisphere gradually emerges from darkness into light.

Neptune

The most distant gas giant, Neptune was discovered in 1846 by Johann Galle in Berlin based on predictions of its existence and location made by John Couch Adams in Cambridge and Urbain Leverrier in Paris due to its effects on the orbit of Uranus. At magnitude 7.8, Neptune is not visible to the unaided eye but can be easily seen through a good pair of binoculars. A telescope is required to see the blueish disc of the planet – its colour due to the methane in the atmosphere absorbing red light rays, as well as blueish icy particles in the clouds above Neptune.

Surface features

The most noticeable feature on the surface was the Great Dark Spot, although recent images by the Hubble Space Telescope show that this has disappeared. Light-coloured cirrus-like clouds of methane ice crystal were also seen in the atmosphere, 50–70km (30-45 miles) above the main cloud layer. The atmospheric layers revolve more slowly at the equator than the poles, taking about 18 hours to complete one revolution. Neptune has three dark, ghostly rings as revealed by the Voyager 2 in 1989. The rings are named after the astronomers Galle, Leverrier and Adams. Galle is the broadest and closest ring.

Opposite left *The Great Dark Spot of Neptune is visible in this image.*

NEPTUNE FACT FILE

❀ **Relation to the Sun:**
   8th planet from the Sun
   Mean distance 4497 million km
   (2794 million miles)

❀ **Relative size:**
   About four times diameter of Earth

❀ **Atmospheric composition:**
   79% hydrogen, 18% helium,
   3% methane, 1% trace gases

❀ **Orbit around the Sun:**
   164.8 Earth years

❀ **Period of rotation:**
   16 hours 7 mins

❀ **Surface temperature:**
   -200°C (-330°F)

❀ **No. of moons:**
   8 moons

Triton

Triton, Neptune's largest and brightest moon (visual magnitude 13.5), has a diameter of 2706km (1682 miles). It orbits Neptune in a retrograde motion – the only large moon to do so in the solar system – at a distance of 355km (220 miles) in a period of 5.9 days.

Photographed by Voyager 2, the satellite has a complex wrinkled surface with no mountains, but cracks of up to 80km (50 miles) wide are evident. The photo (inset, right) shows erupting plumes rising on Triton, which then drift downwind for up to 100km (60 miles).

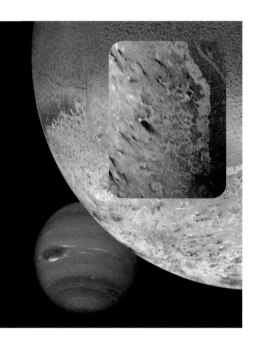

Neptune's largest moon, Triton was discovered soon after the planet. It has a diameter of 2706km (1682 miles) and orbits Neptune in a retrograde motion – the only large moon to do so in the solar system – at a distance of 355km (220 miles) in a period of 5.9 days. Nereid, Neptune's second largest moon was only discovered in 1949, while the remaining six moons were recorded by Voyager 2 in 1989.

Below *This colour image from the Hubble Telescope gives some indication of Neptune's blustery weather bands.*

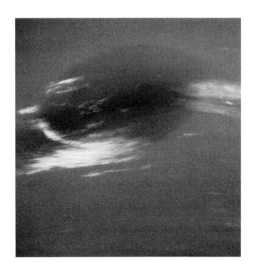

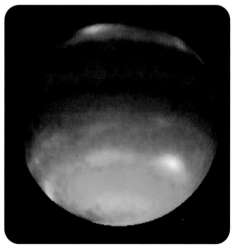

PLUTO FACT FILE

❈ **Relation to the Sun:**
9th planet from the Sun
Mean distance 5900 million km
(3700 million miles)

❈ **Relative size:**
Diameter 2302km (1430 miles)
(smaller than moon)

❈ **Atmospheric composition:**
Nitrogen gas, traces carbon
monoxide, methane and other gases

❈ **Orbit around the Sun:**
248.6 Earth years

❈ **Period of rotation:**
6 days 9 hours

❈ **Surface temperature:**
-225°C (-370°F)

❈ **No. of moons:**
1 moon

Pluto is the smallest planet in our solar system – even smaller than our Moon – with a diameter of 2302km (1430 miles). It was discovered in 1930 by a photographic search made by Clyde Tombaugh at the Lowell Observatory in the USA. Its moon, Charon, was discovered 19000km (11,000 miles) distant from the planet, which it orbits in 6.4 days – the same time Pluto takes to rotate on its own axis. This results in synchronized rotation – each keeps the same face toward the other as they both rotate. Pluto's orbit is quite different to other planets. While other planets travel in near circular orbits close to the elliptical plane, Pluto's orbit takes the planet from around 4400 million km at its closest point to the Sun (perihelion) to 7300 million km (2700 million miles; 4500 million miles) at the most distant point in its orbit (aphelion). When nearest to the Sun it is closer than Neptune!

Observing Pluto

Pluto presents a real challenge. With a magnitude of 13.8, at least an 8in (200mm) telescope located at a very dark observing site is required, coupled with a transparent sky – and you do need to know exactly where to look. In most years, at least one of the astronomical magazines will give a detailed chart showing Pluto's path across the sky, with advice on how to star-hop from a nearby bright star. During the first two decades of this century it will be slowly moving from the southern part of Ophiuchus down through Sagittarius, so it is low in the sky for Northern observers.

Opposite top *This composite photograph of Jupiter and the approaching Comet Shoemaker-Levy 9 was assembled from a number of different images taken by the Hubble Space Telescope.*

Comets

Far beyond the orbit of Pluto lies the Oort Cloud, a region between 1 and 3 light years from the Sun, where, in 1932, Ernst Öpik first suggested that there might be found a reservoir of solar system debris. These are lumps made up of ice and dust – dirty snowballs might be a good description. Called cometary nuclei, they can be many kilometers in size. Sometimes interactions, perhaps with a passing star, can propel them in towards the Sun where, as they reach the inner solar system, the Sun's heat evaporates ice on the surface releasing the dust which forms a curving yellow-white tail along the comet's path. The gases released form a 'coma' surrounding the nucleus and a second, less bright, 'ion' tail, bluish in colour, which lies directly away from the Sun. They can provide some of the most spectacular sights in the heavens!

An eccentric orbit

Pluto's orbit is different from the other planets. While most planets travel in near circular orbits close to the ecliptic plane, Pluto's orbit takes the planet from around 4400 million km at its closest point to the Sun (perihelion) to 7300 million km (2700 million miles; 4500 million miles) at the most distant point in its orbit (aphelion). For a short part of this orbit, most recently between September 1979 to February 1999, it was closer to the Sun than Neptune. Pluto will never collide with Neptune as its orbit is also inclined to the ecliptic plane by 17° and it travels from 1197 million km north of the plane to 1945 million km south (744 million miles; 1209 million miles).

Meteor showers

Each time a short-period comet (one that has been captured to orbit in the solar system) rounds the Sun, dust is released which tends to slowly distribute itself around the orbit – leaving a trail of dust through the solar system. In its own orbit around the Sun, the Earth may cross such a trail and 'sweep up' some of the dust particles, which then fall into the atmosphere. Here they burn up, forming a meteor trail, or shooting star. On almost any dark night one will see a few meteor trails from 'sporadic' meteors.

A few times each year, the Earth crosses a particularly rich stream of dust and we get what is called a meteor shower, seemingly radiating from a point in the sky called 'the radiant'. The name of the shower is given by the constellation in which the radiant lies.

Sometimes the Earth crosses the orbit of the comet close to the comet's actual position. You then have a chance of seeing a meteor storm, when the number of meteors can exceed 1000 per hour! The astronomy magazines will alert you to the possibility of such a spectacular event.

YEARLY METEOR SHOWERS

QUADRANTIDS	4/5	January
	(radiant in Boötes)	
PERSEIDS	11/12	August
LEONIDS	18/19	November
ORIONIDS	21/22	October
GEMINIDS	13/14	December

DISCOVERY OF A 'NEW PLANET'?
In the region beyond Neptune many new bodies – sometimes called Trans-Neptunian Objects (TNOs), sometimes Kuiper Belt Objects – are being discovered. Until recently the largest was Quaoar, about 1290km (800 miles) across, but then, in March 2004, the discovery of Sedna was announced, at possibly 1770km (1100 miles) in diameter – nearly as large as Pluto! Whether this will be regarded as the 10th planet is unlikely; it is more likely that Pluto would be demoted, as it no longer satisfies the generally accepted definition of a planet – which is that the object should be significantly larger than other bodies in its vicinity.

eros

ida

gaspra

dactyl

Left *Photographs of asteroids Eros, Gaspra with it's companion Dactyl, and Ida.*
Opposite *The dramatic Leonids, an annual meteor shower.*

5

THE SKY THROUGH THE YEAR

This chapter provides star charts for the four seasons of the year to help you find your way around the heavens. Each season has two sky maps showing the constellations then seen best by Northern or Southern observers. Their aim is to help you learn the relative positions of the main constellations and thus show only the brighter stars. The text has been divided into four sections, one for each of the four seasons of the year defined by each season's central month (January, April, July and October). A brief description of the sky is provided for each chart, followed by a little more detail on the constellations themselves together with references to the interesting objects within them that are part of the Astronomical A-List and which are described in detail in Chapter 6.

Opposite *Julius Schiller's map of the Christian constellations appears in* The Celestial Atlas, or The Harmony of the Universe *(1661).*

January Sky

Northern Hemisphere Observers

Orion straddles the celestial equator and presents one of the most beautiful star-scapes of the whole year to observers in either hemisphere. Betelgeuse is at its upper left and Rigel to its lower right. Using the three stars of Orion's Belt as a pointer, the constellation Taurus, with its bright orange-red star, Aldebaran, lies up to the right while down to the left is the brilliant white star Sirius in Canis Major. Arcing up toward the zenith from Orion brings you to the yellow star Capella, the brightest star in Auriga, while up to the left of Orion lies Gemini with its 'twin' stars Castor (highest in the sky) and Pollux. Between Gemini and Canis Major lies Canis Minor with its single bright star, Procyon.

Northern and Southern Observers

Orion

Orion is the hunter holding up his club and shield against the charge of Taurus, the bull. Alpha (α) Orionis, or Betelgeuse, its upper left star (lower right star for Southern observers), is a red super-giant varying in size between 300 and 400 times that of the Sun. At Orion's lower right (or upper left) is Beta (β) Orionis, or Rigel, a blue super-giant which is twice the distance of Betelgeuse from the Sun. It has a 7th-magnitude companion. The three stars of Orion's Belt are approximately 1500 light years away. Below the central star of the belt is the 'sword' of Orion, which contains the Orion Nebula, or M42 (❄ p152). M42 is a region where stars are forming out of dust and gas lit up by the light of a group of very hot, young stars at its heart called the Trapezium.

Southern Hemisphere Observers

Perhaps the most spectacular of all constellations, Orion lies (upside down) in the northern sky with Rigel at its upper left and Betelgeuse at its lower right. The stars of the belt point up to the right to Sirius in Canis Major and down to the left to Aldebaran (bright, orange-red) in Taurus. Down to the right of Orion are the twins Castor and Pollux (higher in the sky) in Gemini and, just above the northern horizon, is Capella at the lowest point of Auriga. Procyon in Canis Minor lies up to the right of Pollux and down to the right of Sirius. Overhead is Canopus, the second brightest star in the sky. With transparent skies, the Large Magellanic Cloud, a nearby dwarf irregular galaxy in Dorado, should be seen toward the south from Canopus.

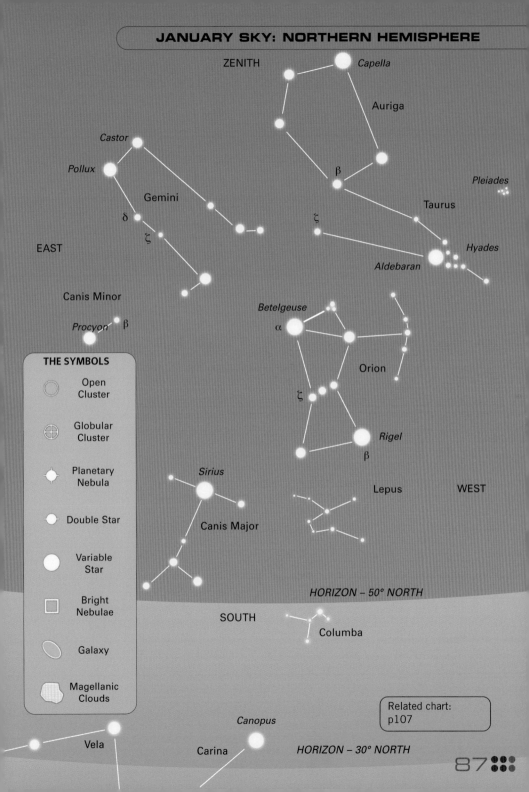

Canis Major

Sirius, the Dog Star, is the brightest star in the sky. This is partly because, with an absolute magnitude of -1.46, it is 20 times brighter than our Sun and partly because it is quite close, only 8.7 light years distant. It has an 8th-magnitude companion, Sirius B, which is a white dwarf star about the size of the Earth but with the same mass as our Sun. This dwarf is very difficult to spot as it is lost in the glare of Sirius. Moving south from Sirius is the open cluster, M41 (✳ p128).

Taurus

Taurus, the Bull, is one of the most beautiful constellations in the heavens. Its face is delineated by the V-shaped cluster of stars called the Hyades (✳ p164), its eye by the orange giant star, Aldebaran, and the tips of his horns by the stars Beta (β) and Zeta (ζ) Tauri. Sweeping one-third of the way from Aldebaran toward Beta (β) Tauri will bring you to an open cluster, NGC 1647, rather overshadowed by its impressive neighbour. A little way from Zeta (ζ) Tauri toward Beta (β) Tauri, can be seen the first entry in Messier's catalogue, M1 – the Crab Nebula – (✳ p167). The jewel of Taurus is the open cluster M45, called the Pleiades (✳ p166). The Pleiades is often called the Seven Sisters, which implies that with your eyes you can see seven stars but, in fact, six stars can normally be seen under clear skies with 10 visible to keen-sighted observers when conditions are ideal.

Gemini

The heads of the twins are marked by the bright stars Castor and Pollux, with magnitudes of 1.6 and 1.1 respectively. Close to the 'foot' of the twin Castor you will, using binoculars, spot the open star cluster, M35 (✳ p144). On the line between Castor and Beta (β) Canis Minoris, and with the same declination as Zeta (ζ) Geminorum, lies the planetary nebula NGC 2392, (✳ p144) called the Eskimo Nebula. It appears as a bluish fuzzy disc about the size of Saturn – which is why such objects are called planetary nebulae! The disc is material thrown off from the 10th-magnitude star at its centre.

Auriga

The hexagonal shape of Auriga represents a charioteer. Its brightest star, Capella, is the sixth brightest star in the heavens. Yellow in colour, it is actually a pair of giant stars 46 light years away from us. Near Capella lies Epsilon (ε) Aurigae which reduces in brightness from 3rd to 4th magnitude every 27 years, remaining at 4th magnitude for 14 months. Astronomers believe that it is then obscured by a disc of dust surrounding its companion star. The next eclipse will begin at the end of 2009. Auriga lies in the plane of the Milky Way, thus one would expect to see signs of star formation in the form of open star clusters such as M36, M37 and M38, which lie within its boundaries.

Dorado

Johann Bayer introduced this constellation, usually said to represent a goldfish, in his star atlas Uranometria, in 1603. However the name comes from the Spanish and actually refers to a mahimahi (dolphin) fish. It has no bright stars but does contain the Large Magellanic Cloud, or LMC (✳ p140), seen as a cloudy patch in the sky on a transparent moonless night. Just separated from the main bar of the galaxy is one of the largest star formation regions that we know of, called the Tarantula Nebula (✳ p140), as the looping arcs

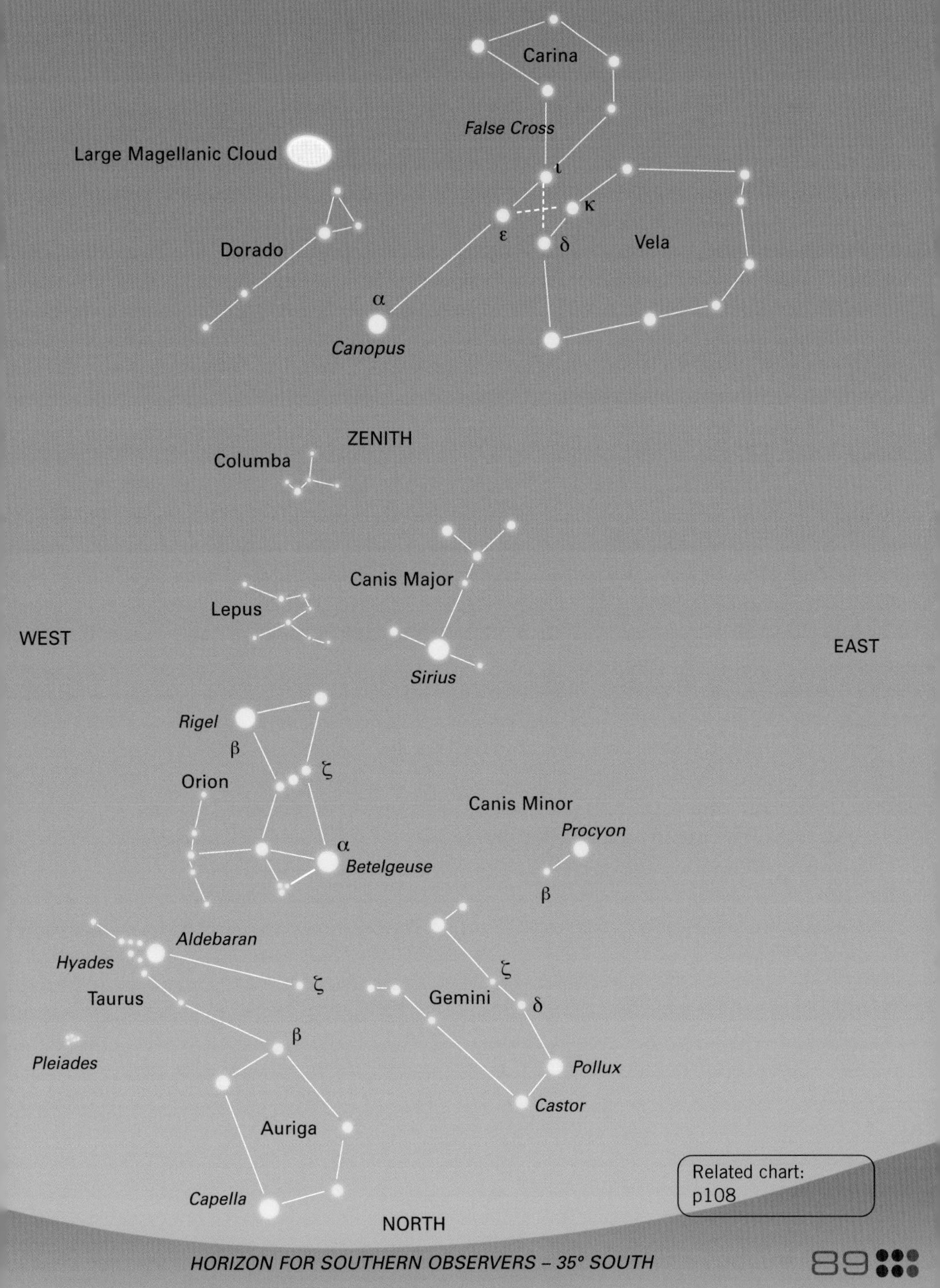

Carina

False Cross

Large Magellanic Cloud

ι

κ

Dorado

ε

δ

Vela

α

Canopus

ZENITH

Columba

Canis Major

Lepus

WEST

EAST

Sirius

Rigel

β

ζ

Orion

Canis Minor

Procyon

α

Betelgeuse

β

Aldebaran

Hyades

ζ

Gemini

ζ

δ

Taurus

Pollux

β

Pleiades

Castor

Auriga

Capella

NORTH

Related chart:
p108

HORIZON FOR SOUTHERN OBSERVERS – 35° SOUTH

89

of gas appear spider-like. The nebula is about 1000 light years across, and were it at the distance of the Orion Nebula, would appear larger than the whole constellation of Orion and be able to cast shadows!

Carina

Carina forms the keel of Argo, the ship that carried Jason on his voyage in search of the Golden Fleece in the Greek myth. In stellar terms, Argo was a huge constellation that has been broken up into four constellations: Carina, the keel; Pyxis, the ship's compass; Puppis, the stern or poop deck; and Vela, the ship's sails. At the western end of the constellation, Canopus (or Alpha [α] Carinae – once Alpha Argus) is the second brightest star in the sky with a magnitude of -0.72. To appear so bright at a distance of 1200 light years means it has to be extremely luminous, which it is – about 200,000 times brighter than our Sun! Two of Carina's stars, Epsilon (ε) and Iota (ι), form part of the False Cross along with two stars of Vela toward the north. Lying almost exactly between the False Cross and the Southern Cross to the east is the Keyhole Nebula – also known as the Eta Carinae Nebula after the roughly 7th-magnitude star Eta (η) Carinae at its heart.

April Sky

Northern Hemisphere Observers

Looking south after dark in springtime will show Gemini and Canis Minor setting in the west and the constellation Leo, including the bright star Regulus, high in the southern sky. Between Leo and Gemini is the faint constellation, Cancer, while down to the left in the southeast, the bright star Spica is seen in Virgo. Just to the north of the zenith

is the Plough, or Big Dipper, forming part of the constellation Ursa Major. The faint constellation Hydra wends its way southward from its head just below Cancer.

Northern and Southern Observers

Leo

Leo is one of the few constellations that resembles its name. Its mane and head form an arc, called the Sickle, with the 1.4-magnitude, blue-white star, Regulus, at its base. Regulus is five times bigger than the Sun, at a distance of 90 light years. Algieba, which forms the base of the neck, resolves into a magnificent double. This pair of golden-yellow giants orbits one another every 600 years. Leo further contains two groups of galaxies that are also well worth observing. Eight degrees greater in right ascension, at the same declination as Regulus, lie the three galaxies M95, M96 (※ p148) and M105 some 27 million light years distant. Close to Theta (θ) Leonis are M65 and M66 (※ p147), appearing as misty patches of light in a small telescope.

Cancer

Though Cancer, the Crab, has no stars greater than magnitude 3.5, it does have a wonderful binocular object in the Beehive Cluster. Also known as Praesepe, the Manger, it is the 44th entry in Messier's catalogue. To the unaided eye, M44 (※ p124) is visible as a misty patch spanning three times the Moon's diameter.

Virgo

Virgo, the maiden holding an ear of wheat, is not a particularly prominent constellation having only

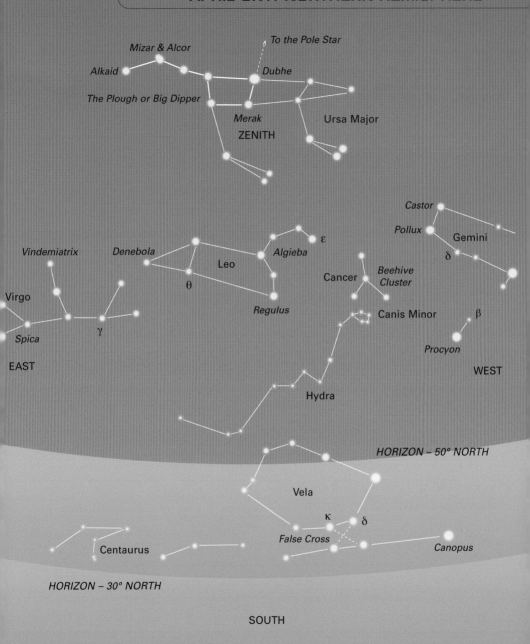

To the Pole Star

Mizar & Alcor

Alkaid

Dubhe

The Plough or Big Dipper

Merak

ZENITH

Ursa Major

Castor

Pollux

Gemini

δ

Vindemiatrix

Denebola

Algieba

ε

Leo

θ

Cancer

Beehive Cluster

β

Virgo

Regulus

Canis Minor

Spica

γ

Procyon

EAST

WEST

Hydra

HORIZON – 50° NORTH

Vela

κ

δ

False Cross

Canopus

Centaurus

HORIZON – 30° NORTH

SOUTH

Related charts:
p109

91

one bright star, 1st-magnitude Spica. The second bright star from Spica toward Leo – Gamma (γ) Virginis, or Porrima (❋ p176) – is a double star. In the region of Virgo close to Denebola, the 'tail' of Leo, lies the centre of the Virgo cluster of galaxies. There are 13 galaxies in the Messier catalogue found in this region, all of which can be seen with a small telescope under dark and transparent skies. The brightest is M87 (❋ p174), which lies on the line between Denebola and Vindemiatrix, in Virgo. Toward the constellation Corvus lies the edge-on spiral galaxy M104 (❋ p176), also called the Sombrero Galaxy, which has a prominent dust lane across its heart.

Northern Observers Only

Ursa Major

The stars of the Plough form one of the most recognized star patterns in the Northern sky. Also called the Big Dipper, after the soup ladles used by farmers' wives in America to serve soup to the farm workers, it forms part of the Ursa Major, or Great Bear, constellation – not quite so easy to make out! The stars Merak and Dubhe form the pointers which will lead you to the Pole Star (Polaris), and hence, North. The stars Alcor and Mizar (❋ p170) form a naked-eye double which is well worth observing in a small telescope as Mizar is then shown to be an easily resolved double star. Ursa Major also contains interesting deep-sky objects. In the upper right of the constellation is a pair of interacting galaxies M81 and M82 (❋ p170). Another galaxy, M101, looks like a pinwheel firework, hence its other name, the Pinwheel Galaxy. It forms a triangle above the two leftmost stars of the Plough's handle. From Mizar, follow the line of 5th- to 6th-magnitude stars toward it. It is a type Sc spiral galaxy seen face-on at a distance of about 24 million light years. Such galaxies have a relatively small nucleus and open spiral arms. A fur-

ther beautiful galaxy, M51 (❋ p172), though just outside the constellation boundary, lies close to Alkaid, the leftmost star of the Plough.

April Sky

Southern Observers

Leo, lying on its back, is in the north, with Castor and Pollux in Gemini setting in the west down to its left. Between Leo and Gemini is the faint constellation Cancer, while up to the right of Leo lies Spica, the only bright star in Virgo. The Milky Way arcs south of the zenith. In the Milky Way, and just to the southeast of the zenith, is the small constellation Crux, the Southern Cross. Looking a little northeast of Crux, and surrounding it on three sides, is Centaurus. Its two very bright stars Alpha (α) and Beta (β) Centauri act as pointers to Crux. Four stars to the west of the zenith form a second cross called the False Cross – this is often confused with the smaller Southern Cross.

Southern Observers Only

Crux – the Southern Cross

Acrux (or Alpha [α] Crucis) (❋ p134), the brightest of the four stars that form the cross, is a triple system of blue stars 500 light years away. Close to the most easterly star of the cross, Beta (β) Crucis (Becrux), lies NGC 4755, or C94. It is also called the Jewel Box (❋ p136) because the brighter 6th- to 8th-magnitude stars appear like 'a casket of variously coloured precious stones', to quote John Herschel. Just south of Beta Crucis is a region that is far darker than the surrounding

APRIL SKY: SOUTHERN HEMISPHERE

Coal Sack

α

β

Carina

α

β

Crux

Centaurus

ε

ι

Canopus

False Cross

δ

κ

Vela

ZENITH

WEST

EAST

Hydra

Procyon

β

Spica

Canis Minor

Regulus

Virgo

Beehive Cluster

Leo

Cancer

Denebola

Gemini

δ

ε

Algieba

Pollux

Castor

Ursa Major

Merak

HORIZON – 35° SOUTH

NORTH

Related charts:
p113

Milky Way. Called the Coal Sack, C99 (�֍ p136), it is a dark nebula over 7° across, formed of dust which is obscuring the more distant stars. It makes a good binocular object.

Centaurus

Alpha Centauri (✷ p132), a triple star system, is just 4.4 light years away. It is made up of two bright stars, A and B, together with an 11th-magnitude companion, C, nearly 2° away in position. At present C is the closest star to the Sun and is thus called Proxima Centauri. With a magnitude of 10, Proxima is a red dwarf star about one-tenth the mass of the Sun, having a diameter of only 200,000km (125,000 miles). It is called a flare star because roughly every six months or so its brightness increases by up to a magnitude as giant flares erupt from its surface. Centaurus contains two A-List objects: Omega (ω) Centauri, (✷ p132), the finest globular cluster in the heavens and the 7th-magnitude galaxy NGC 5128 whose common name is Centaurus A (✷ p133) as it is one of the strongest radio sources in the sky.

July Sky

Northern Observers

The star Vega, in the Lyra, is almost overhead, with Cygnus, the Swan, a little toward the east with its brightest star, Deneb (Alpha [α] Cygni), forming its tail. Its central section is also called the Northern Cross. Down to the southeast is the bright star Altair in Aquila. Vega, Deneb and Altair form what is called the Summer Triangle. Arcturus, in Boötes, lies high in the west, and the line from Vega to Arcturus passes through Hercules and Corona Borealis. Low in the south are Sagittarius and Scorpius, the latter with its prominent bright star, Antares.

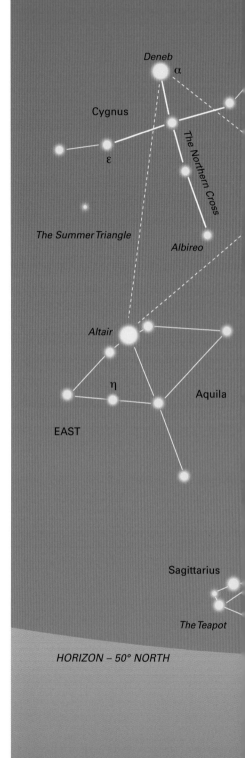

Deneb
α
Cygnus
The Northern Cross
ε
Albireo
The Summer Triangle
Altair
η
Aquila
EAST
Sagittarius
The Teapot
HORIZON – 50° NORTH

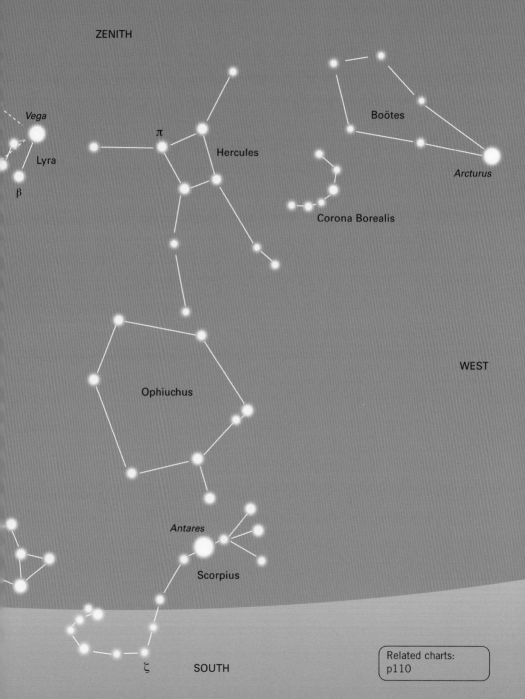

ZENITH

Vega

Lyra

β

π

Hercules

Boötes

Arcturus

Corona Borealis

Ophiuchus

WEST

Antares

Scorpius

ζ

SOUTH

Related charts:
p110

Lyra

Lyra is dominated by its brightest star Vega, the fifth brightest in the sky. It is a blue-white star having a magnitude of 0.03, and lies 26 light years away. It is three times the mass of the Sun and is about 50 times brighter – thus burning up its nuclear fuel at a greater rate and so will shine for a correspondingly shorter time. Vega is much younger than the Sun, perhaps only a few hundred million years old, and is surrounded by a cold, dark disc of dust in which an embryonic planetary system is being formed.

There is a wonderful multiple star system called Epsilon (ε) Lyrae (✳ p150) up and to the left of Vega whilst between Beta (β) and Gamma (γ) Lyrae lies the Ring Nebula – the 57th object in the Messier Catalogue, (✳ p152).

Cygnus and Vulpecula

Cygnus, the Swan, is sometimes called the Northern Cross as it has a distinctive cross shape, but we normally think of it as a flying swan. Deneb, the Arabic word for 'tail', is a 1.3-magnitude star that marks the tail of the Swan. It is nearly 2000 light years away and appears so bright only because it gives out around 80,000 times as much light as the Sun.

The star, Albireo (✳ p136), which marks the head of the Swan, is much fainter but through a small telescope you see that Albireo is made up of two stars, amber and blue-green, which together provide a wonderful colour contrast. South of the star marking the left-hand arm of the Northern Cross, Epsilon (ε) Cygni, lies the

supernova remnant of a star that exploded some 15,000 years ago. The brightest part to observe under dark and transparent skies is called the Veil Nebula, C34 (✳ p138). Cygnus lies along the line of the MilkyWay, and provides a wealth of stars and clusters to observe. On a dark transparent night, a dark lane in the Milky Way, called the Cygnus Rift, is seen to the side of Cygnus toward Altair. It is caused by the obscuration of light from distant stars by a lane of dust in our local spiral arm. In this rift and in the constellation Vulpecula, on the line between Vega and Altair, is a rather pretty asterism (a chance pattern of stars), properly called Brocci's Cluster but usually known as the Coathanger (✳ p178) as that is exactly what it looks like through binoculars. Vulpecula also contains a planetary nebula, M27 – the Dumbbell Nebula, (✳ p178), well worth searching out.

WEST

Arcturus

July Sky

Southern Observers

Overhead, within the arc of the Milky Way, lie Sagittarius and Scorpius, the latter with its bright star, Antares. Following the Milky Way

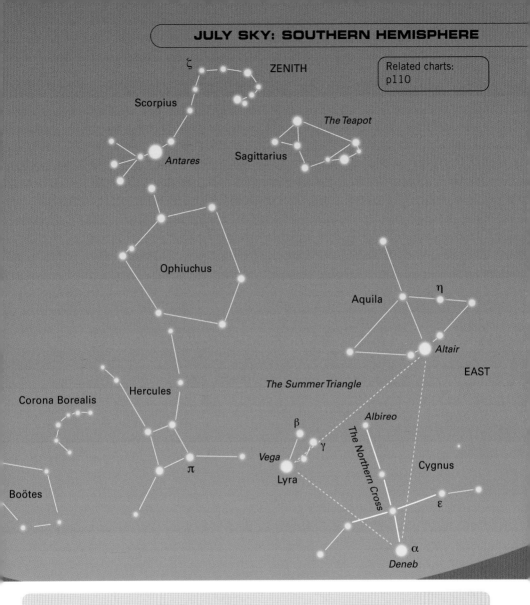

Related charts:
p110

toward the northeast is the bright star Altair in Aquila. This is the highest star in the Summer Triangle, formed with Vega in Lyra down to the left, and Deneb in Cygnus close to the horizon down to Vega's right.

Above the northwestern horizon is the bright star Arcturus in the constellation Boötes. Lying between Arcturus and Lyra are, first, Corona Borealis and then the keystone of Hercules.

Aquila

Aquila represents an Eagle. Its brightest star, Altair, is one of the 20 brightest stars in the sky and quite close to us at a distance of just under 17 light years. Eta (η) Aquilae (❋ p122) was the first Cepheid variable to be discovered – by Edward Piggot in 1784. Cepheid variables play an important role in measuring distances to galaxies.

Hercules

The body of Hercules is made up of four stars that form a 'keystone'. Three are of magnitude 3.5, the fourth 3.9. Two-thirds of the way northwards up the western arm is M13 (❋ p145), the most spectacular globular cluster in the northern sky. North of the keystone and 6° due north of Pi (π) Herculis is a second bright globular cluster, M92 (❋ p146).

Sagittarius and Scorpius

The body of Sagittarius, in the form of a 'teapot' (which is upside down for Southern observers), lies in the plane of the Milky Way toward the galactic centre. Two deep-sky objects appear reasonably prominently in binoculars. If observers follow the 'flow' of water southward from the spout, a misty patch can be seen, the open cluster M7 (❋ p162), just into the constellation Scorpius. To the north-west of M7 is a second open cluster, M6 (❋ p162). Close to the star Zeta (ζ) Scorpii southern observers will be able to see a beautiful and very rich region of the Milky Way. At its heart is a cluster, NGC 6231 (❋ p164) often called the Northern Jewel Box. The region is sometimes called the False Comet, with the 'jewel box' as the coma, and the nebulosity and rich star fields above forming the tail. Above the 'lid' of the 'teapot' (below the lid for Southern observers) is a fainter region of the Milky Way that contains a bright nebulosity, M8 (❋ p159), also called the Lagoon Nebula. It is easily visible on a dark transparent night with binoculars. Averted vision will help you see the full extent of the nebulosity. Not far from M8 lies M20, the Trifid Nebula (❋ p160) which has a complex structure of dark dust lanes dividing the nebulosity into three – hence its name. North-east of M8, in the direction of Altair, is M17 (❋ p160), the Omega (ω), or Swan, Nebula – looking rather like the body and curving neck of a swan.

Perseus

October Sky

Northern Observers

The summer triangle of Deneb, Vega and Altair is now seen in the western sky with the square of Pegasus to the southeast. Up to the left of Pegasus, viewed on its back by Northern observers, is the arc of the constellation Andromeda, while close to the zenith is the W-shape of Cassiopeia lying in the Milky Way. Down along the band of the Milky Way, to the east of Cassiopeia, is Perseus.

EAS

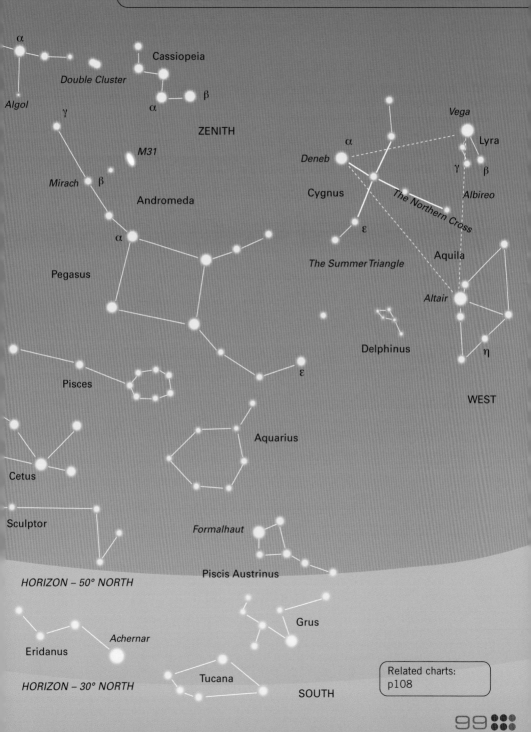

α
Cassiopeia
Double Cluster
β
Algol
α
γ
ZENITH
Vega
Lyra
M31
α
Deneb
γ
β
Mirach
β
Cygnus
Albireo
Andromeda
The Northern Cross
α
ε
Aquila
Pegasus
The Summer Triangle
Altair
Delphinus
η
ε
Pisces
WEST
Aquarius
Cetus
Sculptor
Formalhaut
Piscis Austrinus
HORIZON – 50° NORTH
Grus
Eridanus
Achernar
Tucana
HORIZON – 30° NORTH
SOUTH

Related charts:
p108

Cassiopeia

This constellation has a distinctive W-shape and is one of the easiest to find in the sky. Start with the pointers in the Plough and follow a curving convex line through the Pole Star until you reach Cassiopeia, lying in a rich part of the Milky Way and containing many star clusters. M52 is the brightest of these, seen as a fuzzy patch in binoculars along a line, from Alpha (α) through Beta (β) Cassiopeiae (the two brightest stars), extending by the same distance as their separation. A small telescope will show a pretty field of stars containing an 8th-magnitude orange giant, appearing brighter than the rest.

Though not visible to small optical telescopes, Cassiopeia also contains two interesting radio sources. Cassiopeia A, the strongest radio source in the sky, is thought to be the remnant of a supernova which exploded in the 17th century but which was never observed at the time. The second is the remnant of what is called Tycho's supernova (see p13), which exploded in 1572 and was visible to the unaided eye for over a year – at its peak it was as bright as Venus.

Perseus

Lying between Cassiopeia and Perseus, the unaided eye might perceive a slightly brighter fuzzy patch in the richness of the Milky Way. Binoculars will make it more distinct but it is best seen with a telescope at low power, when both of the two clusters that make up the double cluster, C14 (❉ p156), can be seen in the same field. The lower bright star in Perseus, forming a right-angled triangle with Alpha (α) Persei and Gamma (γ) Andromedai, is Algol (❉ p158). It is called the 'demon star' as it appears to wink precisely every 2.87 days when its brightness falls by 1 magnitude for two hours. In fact, Algol is an eclipsing binary system (see p31) with a red subgiant star eclipsing the brighter blue giant.

October Sky

Southern Observers

In the north lies the square of Pegasus with Andromeda arcing down to the northeastern horizon. Altair, in Aquila, will be setting toward the west with Deneb, in Cygnus, just above the northwestern horizon. At the zenith, to the west of the bright star Achernar, is the constellation Tucana. Still in Tucana, toward the south celestial pole, is the misty patch of the Small Magellanic Cloud (SMC).

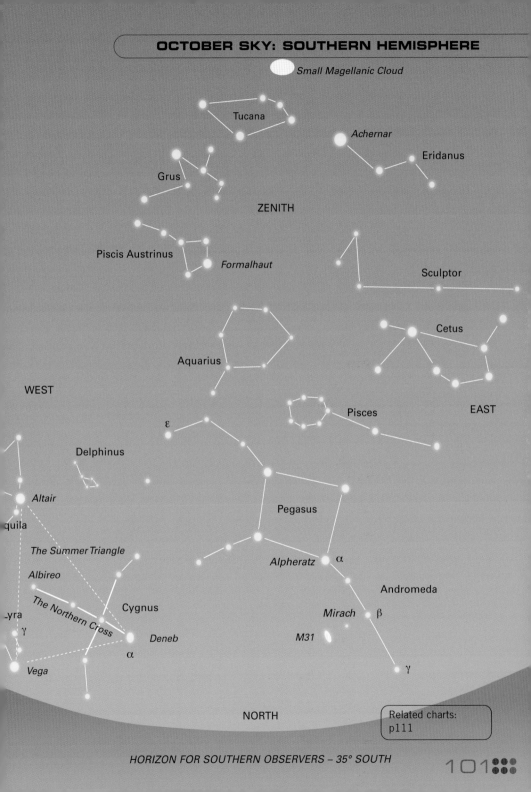

Pegasus

The number of stars visible within the square of Pegasus, the Winged Horse, is a good test of the sky's transparency. If you can see four stars, it's pretty good! Just to the west of the western side of the square is the 6th-magnitude star, 51 Pegasi (✳ p156). Looking at this star, try to imagine a planet half Jupiter's mass orbiting it once every 4.2 days. It was the first planet (unnamed) to be discovered in orbit around a star like our Sun.

The highlight of this constellation lies a short way up to the right (down to the left for Southern observers) of Epsilon (ε) Pegasi, the nose of the horse. There, binoculars will show a small fuzzy patch that is the globular cluster, M15 (✳ p155).

Andromeda

The top left star of the square of Pegasus (for Northern observers; bottom right for Southern), 2.5-magnitude Alpheratz (Alpha [α] Andromeda) is actually in Andromeda. North-east of this star lies a hazy oval patch of light that is the nucleus of M31 (✳ p120), the Great Nebula in the Andromeda. The mutual gravitational attraction between M31 and our own Milky Way is bringing them together so that, in a few billion years, they may well merge into one. To the south-east of M31, in neighboring Triangulum, binoculars might show another misty patch – but only in near perfect conditions with no Moon. This is M33 (✳ p122), the third largest spiral galaxy in our local group. Details of how to find M31 and M33 are given in the next chapter.

Tucana

The Large and Small Magellanic Clouds lie toward the south celestial pole, roughly alongside a line between the bright stars Achernar, high in the sky, and Canopus, toward the southeastern horizon. The LMC (✳ p140) is to the east of this line with the SMC toward the south. The unaided eye and binoculars will show them well in moonless skies. Like the LMC, the SMC is a satellite galaxy of our own Milky Way. Close to the SMC, and easily visible in binoculars, lies a globular cluster of 5th magnitude called 47 Tucanae (NGC 104), (✳ p169). It has virtually the same brightness as Omega (ω) Centauri, the other spectacular Southern globular cluster.

Right *Although the Andromeda Nebula is the nearest galaxy to our own Milky Way, it is now thought to be 2.9 million light years distant.*

THE ALL
SKY
CHARTS

This section comprises a set of star charts covering the whole sky; two for the regions around the north and south celestial poles and a further six to circle the sky around the celestial equator. Stars down to the unaided eye limit, 6th magnitude, are plotted on a grid showing the celestial co-ordinates in Right Ascension and Declination. The names of the brightest stars are given along with the names and outline shapes of the major constellations. Also shown is the path of the Sun across the sky, called the Ecliptic. The locations and names of the deep-sky objects described in the book are plotted on the chart, with each represented by a symbol which indicates what type of object it is. The key to the symbols used on page 87 also applies to the all-sky charts.

Left *The Kitt Peak Observatory in Arizona, USA, is home to a number of telescopes – 4m, 2.1m and 0.9m, among others. Its facilities also include a solar observatory.*

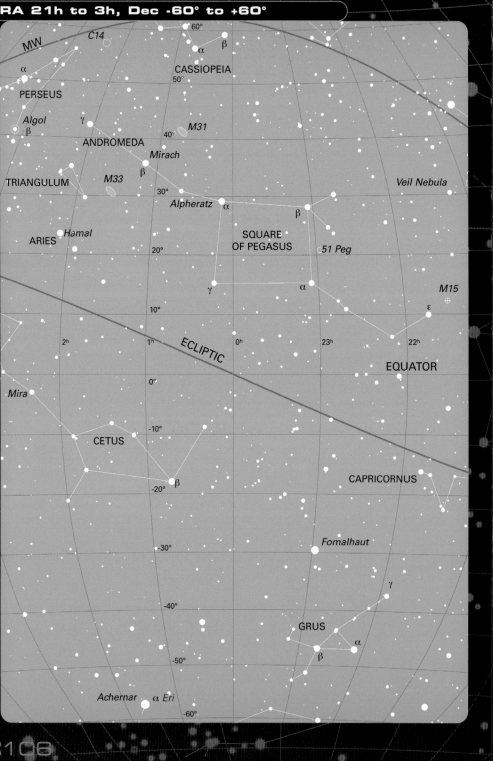

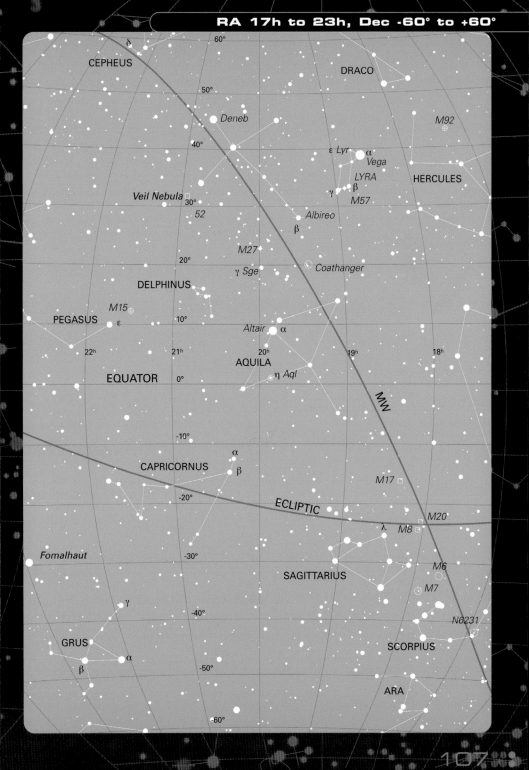

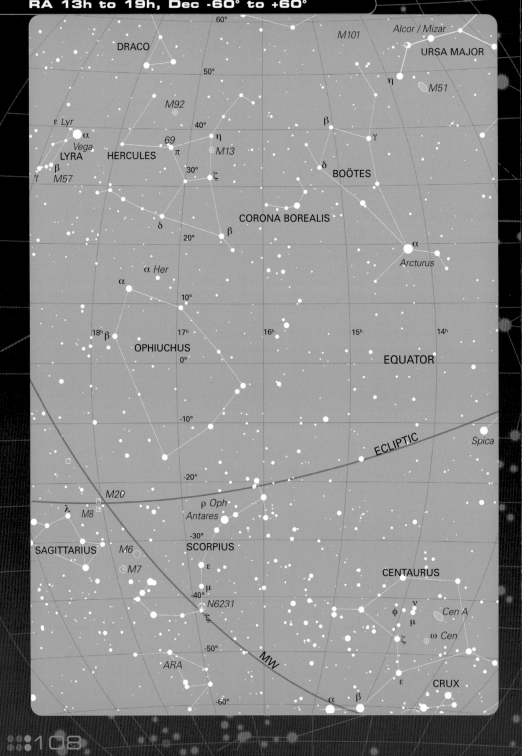

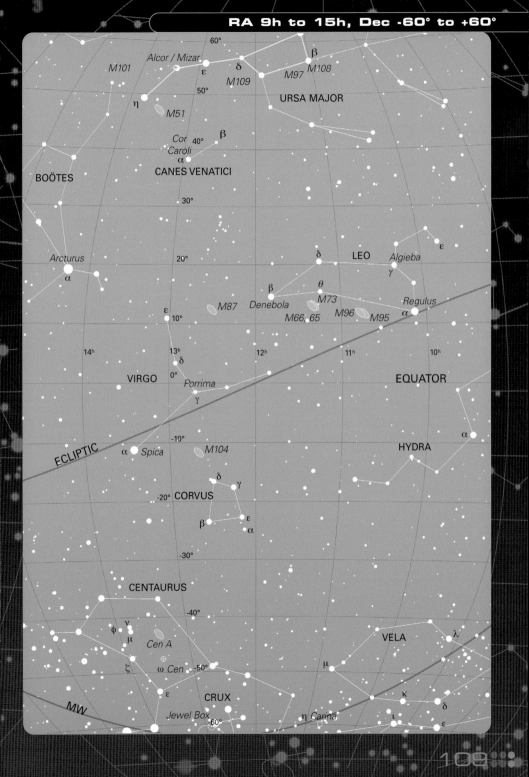

60°

Alcor / Mizar

M101

ε δ β

M109 M97 M108

50°

η M51 URSA MAJOR

Cor 40° β
Caroli
α

BOÖTES CANES VENATICI

30°

20° δ LEO Algieba

Arcturus β θ γ

α ε M87 Denebola M73 Regulus
10° M66, 65 M96 M95 α

14h 13h 12h 11h 10h

δ
VIRGO 0° Porrima EQUATOR
γ

-10° M104 HYDRA
α Spica

δ γ α
-20° CORVUS

β ε
α

-30°

CENTAURUS

-40°

ν
φ λ
μ VELA
Cen A
ζ μ
ω Cen -50° κ
δ
ε CRUX ι
ECLIPTIC Jewel Box η Carina ε
MW -60°

109

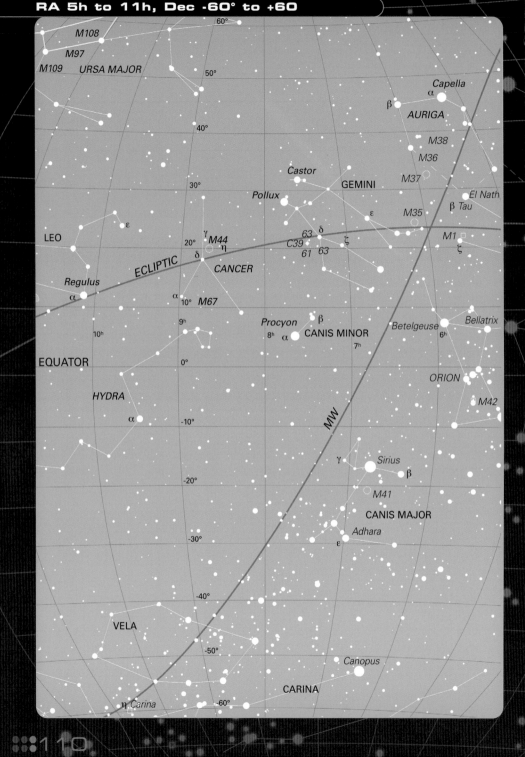

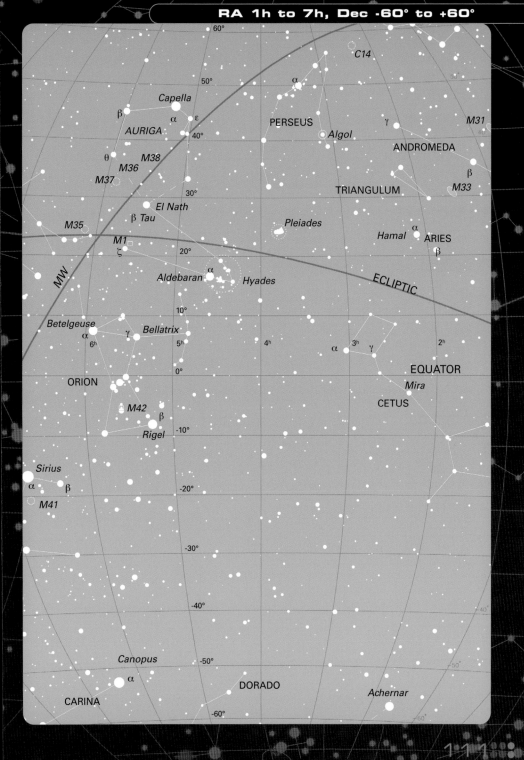

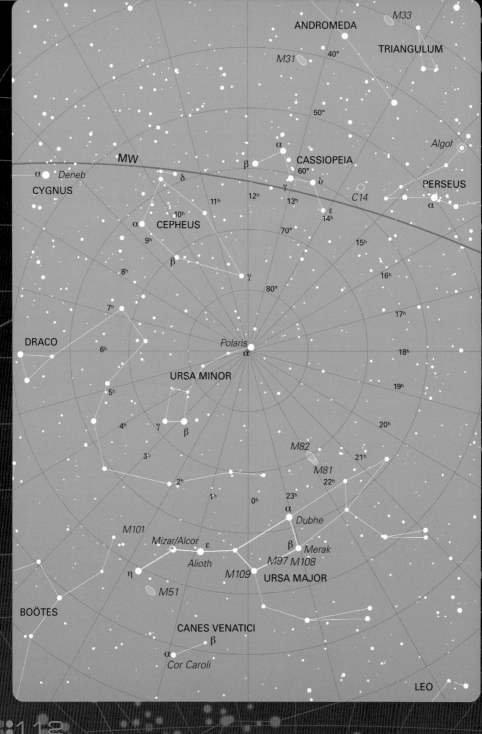

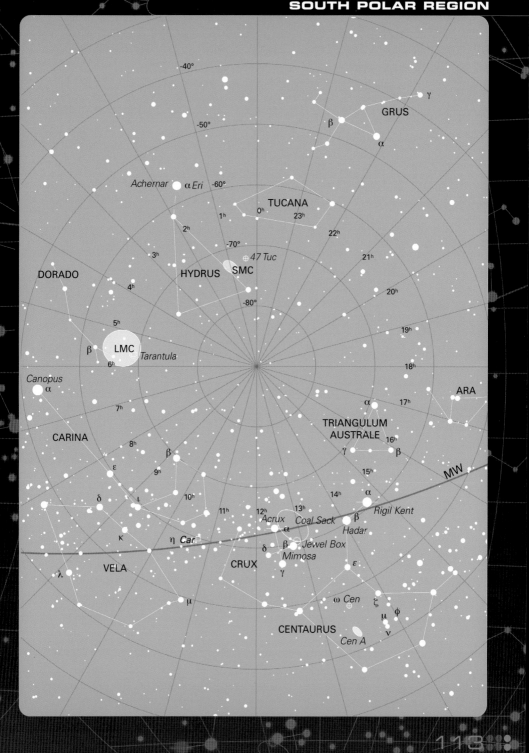

6

OBSERVING THE ASTRONOMICAL A-LIST

This chapter will provide you with guidance for observing 50 of the best objects in the sky, as first introduced in the 'Astronomical A-List'. Many are visible to the unaided eye, most with binoculars and all with a small telescope. There are many catalogues of celestial objects, the most relevant to amateurs being the Messier Catalogue produced by Charles Messier to provide a catalogue of diffuse objects that might be confused with comets. It contains many of the best objects to observe in the northern sky. But, as Messier was observing from Paris, it cannot include the objects in the southern skies. The Caldwell Catalogue, compiled by Patrick Moore, does include Southern sky objects but neither includes individual star systems. However, the A-List is totally inclusive so that if an amateur has observed the majority, an example of almost every type of interesting celestial object will have been seen.

Opposite *The Eagle Nebula is filled with actively forming stars; it has an embedded star cluster at its heart. The name derives from the dust columns outlined against bright clouds of gas.*
Previous pages *Arecibo's (Puerto Rico) spherical reflector dish has a 305m (1000ft) diameter.*

Why 50 objects?

The list tries to include all of the different types of objects – but only the best. The Messier and Caldwell catalogues include many globular clusters but perhaps only five stand out: M13, M92 and M15 in the northern sky and, even better, Omega Centauri and 47 Tucanae in the southern sky. All these have been included. The same reasoning was applied to the other types of objects too, and the list ended up at 50. Some entries are pairs of objects that can be seen together in a single telescope field of view such as the galaxies M81 and M82. So the actual number of individual objects is actually higher than 50.

Here are some statistics. Twenty-five of the objects are in Messier's Catalogue and twelve are in the Caldwell Catalogue. This leaves 13 that occur in neither. These are mostly star systems that are well worth observing, such as Algol – the 'demon' star that 'winks' due to its occultation by its companion star every 2.867 days.

Note: We've included a table on pp190–91, which lists the Astronomcal A-list objects.

Each entry gives the Messier or Caldwell number, the common name (if any) and the type of object. Following this is a sequence of bold letters. These indicate how the object may be viewed:

E	The unaided eye
B	Binoculars
L	Telescope at low power
M	Telescope at medium power
H	Telescope at high power

The positions of the object are for equinox 2000.

Above *M92 is a globular cluster that can be located in the sky using a pair of binoculars.*

Best time to find the objects

Some of these objects can be seen whenever they are reasonably high in a clear sky. Other objects, particularly galaxies, will require dark and transparent skies to be seen well. What does this mean?

A **dark sky** is one when the Moon is not too bright (a thin crescent will not matter, a Full Moon will) and there is not too much light pollution – that is, away from towns or cities.

A **transparent sky** is one where there is little water vapour or dust in the atmosphere to scatter light. Not only do both of these reduce the apparent brightness of objects but they also reflect any light pollution, making their effect even worse. Nights following heavy rain are often ideal as the dust has been washed out of the atmosphere. Other objects, such as close binary stars, need a night of good seeing when the atmosphere is steady and the stars do not twinkle too much. Not surprisingly, objects look at their best on nights when the atmosphere is both transparent and steady but, sadly, these don't occur too often.

How to find these objects in the night sky

The introduction to each constellation indicates when the objects within them can be seen at their best. This is simply when they are highest in the sky so that you are observing through the minimum amount of atmosphere. The charts will show you where the objects are in relation to the brighter stars of the constellation. In these charts north is to the top, south to the bottom, east to the left and west to the right. In some cases, such as the Pleiades cluster, no further instructions are needed, in others the object may be very close to a bright star and so can be located easily.

Where the object under observation is away from any bright stars, there are three methods that can be used to find them:

1. **Star hopping** One starts with an obvious bright star and follows a series of instructions to 'hop' from one star to another to find the object in question.

2. **The geometrical method** The object may make, for example, a right angle or equilateral triangle with two other stars so, having found these stars, perhaps by star hopping, one can visualize where the object should lie in the sky and point the telescope towards that region.

3. **Scanning in Right Ascension (RA) or Declination (Dec)** – this requires an equatorially mounted telescope. Suppose an object is either precisely north or south of an obvious star. If this star is first centred in the finder scope or telescope field of view and the RA axis locked, moving the telescope in Dec, up or down as appropriate, will bring you to the desired object. By using the Dec setting circle, it is often possible to offset the telescope pointing by the correct number of degrees so it will immediately appear in the field of view using a low-power eyepiece. Just the same technique will bring you to objects that have the same Dec as an obvious bright star. In this case you can set the RA setting circle to zero, so reading off the offset directly – note that the RA circle is graduated in minutes of time not angle; 1 hour is equivalent to 15° at the equator, 4 minutes of time = 1°. (When an object is on or close to the meridian – the north-south line – RA and Dec equate to Azimuth and Altitude respectively, so this technique can also be used with an Alt/Az telescope mount.)

The information about each object will indicate how you can best find them using the techniques described above.

> **A word of warning** The details provided about each of the objects contain many facts such as brightness, size, distance and age. These are not always well known and different sources give different values. The text indicates when a fact is not known accurately and gives the values that are believed to be most accurate. Please do not be too surprised if you come across differing values in other books or on the Internet!

One aim of the A-list is for it to be used to encourage amateur astronomers, especially youngsters, to make their own observations to gain Bronze, Silver and Gold observing awards by submitting logbooks of their observations by post or e-mail. The certificates will be awarded jointly by the University of Manchester's Jodrell Bank Observatory and the UK's Society for Popular Astronomy.

Full details on how to submit observing logs for the awards can be found at the following website: www.jb.man.ac.uk/public/Alist.html

The A-List

The Constellations

Andromeda and Triangulum

These adjacent constellations contain two A-List objects: the galaxies M31 and M33. These, along with our galaxy, the Milky Way, are the three major galaxies in our Local Group of galaxies. The distribution of galaxies in the Local Group is shaped a little like a dumbbell with the Milky Way galaxy at the centre of one 'weight' with M31 and M33 at the heart of the other. This is why they appear close in the sky. These galaxies are best seen during the months before Christmas, high overhead for Northern observers, but low in the north for Southern skywatchers.

M31   Andromeda

The Andromeda Galaxy
• **Spiral galaxy**   **E B L**

M31 is the nearest large galaxy to us lying at a distance of 2.9 million light years. It is the largest galaxy in our Local Group (*see* p42) and

is a spiral galaxy of type Sb, somewhat larger than our own Milky Way. (Type Sa spirals have a large nucleus and very tight spiral arms, type Sc have a small compact nucleus with very open arms. Type Sb lies in between.) With a magnitude of 3.4, M31 is visible to the unaided eye – the most distant object in the universe that most people can see using just their eyes! The easiest way to locate it is to star hop from the star Alpheratz, α Andromedae, which forms the north-western star of the square of Pegasus. First move two stars eastwards to the star Mirach, then turn 90° clockwise and move to the first bright star. Moving in the same direction by the same distance, you should then easily pick up a fuzzy white glow that is the nucleus of M31. If Pegasus is low in the sky, a second way of finding it is to follow the 'arrow' made by the west-most three stars of Cassiopeia. Andromeda is 15° (about three binocular fields of view) from the tip of the arrow.

You can see the heart of M31 with your un-aided eye as a fuzzy white patch of light, but it is best seen with binoculars under dark and transparent skies. Then, with well dark-adapted eyes, it is possible to get some feeling of the extent of the galaxy which stretches some 3 by 1° in angular size – spanning half way across a typical binocular field. Using a telescope with a low-power eyepiece is also rewarding because, if one slowly sweeps the field of view across the galaxy under dark skies, it is possible to pick out some of the dark dust lanes that cut across the faint starlight.

M31 has two daughter galaxies M32 and M110. Both are elliptical and seen as tight fuzzy balls. M32, magnitude 8.4, appears closer to the centre of M31 and is a type E2 galaxy – almost spherical. M110, of type E6 and magnitude 8.5, is more obviously elongated.

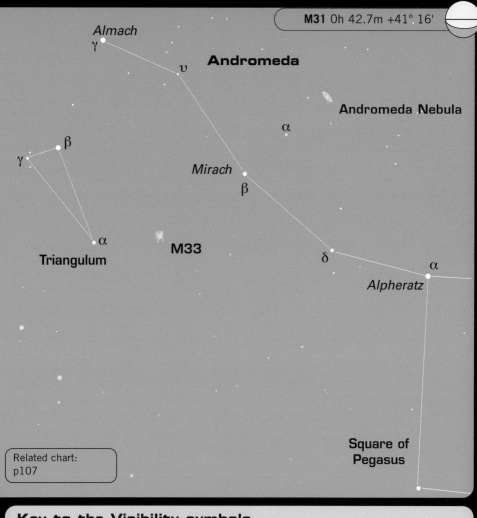

M31 0h 42.7m +41° 16'

Almach
γ

υ

Andromeda

Andromeda Nebula

α

Mirach
β

β

M33

α

δ

α

Alpheratz

Triangulum

γ

Square of Pegasus

Related chart:
p107

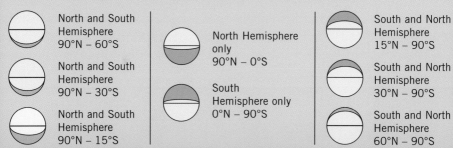

Key to the Visibility symbols

Please note: These are a general indication as to what stars are visible from each hemisphere - the degrees are averages as they vary from season to season.

North and South Hemisphere 90°N – 60°S

North and South Hemisphere 90°N – 30°S

North and South Hemisphere 90°N – 15°S

North Hemisphere only 90°N – 0°S

South Hemisphere only 0°N – 90°S

South and North Hemisphere 15°N – 90°S

South and North Hemisphere 30°N – 90°S

South and North Hemisphere 60°N – 90°S

The Triangulum Galaxy
• Spiral galaxy   **B L**

M33 is a face-on type Sc spiral galaxy having a small compact nucleus and open spiral arms. It lies at a distance of 3 million light years and is the third largest galaxy in our Local Group. With a magnitude of 5.7, those with very acute eyesight under perfect conditions might just be able to observe it with the unaided eye but, as it is face-on to us, the light is well spread out and most will only ever observe it with binoculars or a telescope – and even with such optical aids, very dark and transparent skies are needed.

From M31, retrace your step to the star Mirach, and continue on in the same direction for about the same distance – 7° or between one and two binocular fields. In binoculars, it appears like a little piece of tissue paper stuck on the sky – just a little brighter than the surroundings. Larger telescopes – say 8in (200mm) or more – will just show a hint of the open spiral arms surrounding the compact nucleus. With an overall diameter of about 60,000 light years, M33 is much smaller than M31 or our own Milky Way, but it is more typical of spiral galaxies across the universe.

M33: 01h 33.9m +30° 39'

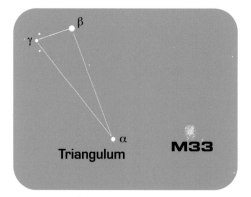

Triangulum   M33

The constellation
Aquila

Altair, Aquila's brightest star is the most southerly of the three stars that make up the summer triangle (Deneb in Cygnus, Vega in Lyra and Altair – *see* the Vulpecula star chart on p179) so, not surprisingly, Aquila is best observed in August, September and October. It contains one A-list object, Eta (η) Aquila – the first Cepheid variable star to have been discovered.

Aquila

Eta (α) Aquilae • Cepheid
variable star   **E B M**

Eta Aquilae lies 8° (one and a half binocular fields) directly south of Altair. It is a bright Cepheid variable that varies in magnitude from 3.7 to 4.5 with a period of 7.2 days. Its variability was discovered by the English astronomer Edward Piggot in 1784. Try to observe it over a period of time and compare its brightness with Iota (ι) Aquilae 4° away. (For Northern observers, put (η) Aquilae at the upper left of a binocular field and (ι) Aquilae will be at the lower right – the opposite for Southern observers). Iota Aquilae has a magnitude of 4.36, so at minimum (η) Aquilae will be slightly less bright than ι (but probably not noticeably so), but at maximum it will be obviously brighter.

Soon afterwards, Piggot's deaf mute neighbour, John Goodricke, discovered a second star called δ Cephei, which varied in a similar way. These stars, which are some of the intrinsically brightest in the heavens, are unstable and oscillate in size and brightness with a very regular period. They became known as Cepheid variables – after δ Cepheus, the second such star to be

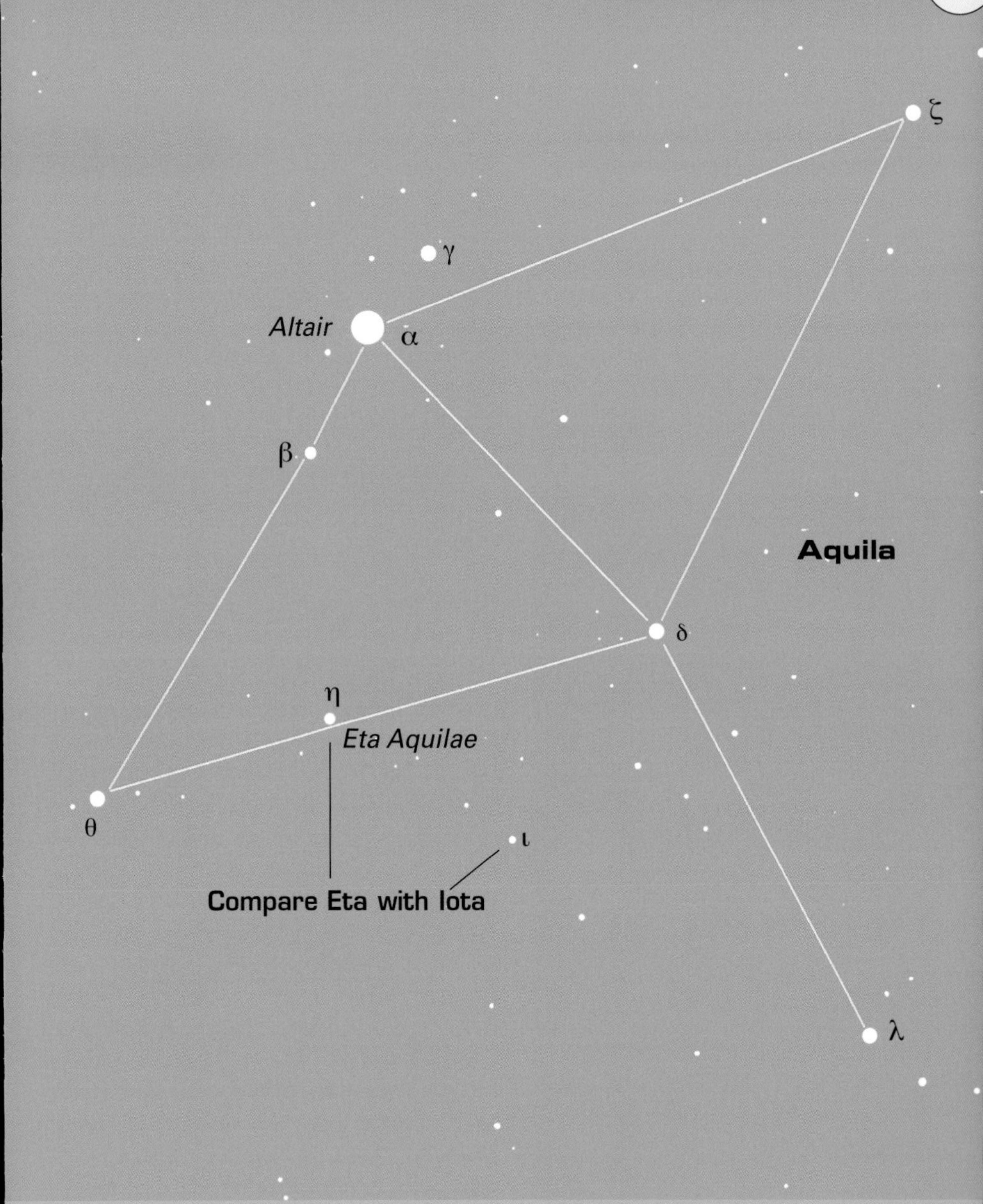

ζ

γ

Altair

α

β

Aquila

δ

η

Eta Aquilae

ι

θ

Compare Eta with Iota

λ

Related chart:
p111

discovered. They have played an important role in measuring the size of the universe. Early in the last century an American astronomer, Henrietta Leavitt, discovered that the brightness of these stars varied with a period that was in proportion to their absolute brightness (*see* p31). As they are very bright, they could be seen in quite distant galaxies. So by simply measuring their periods, their absolute brightness can be found and hence the distance to the galaxy in which they are located. Observations of Cepheid variables in remote galaxies by the Hubble Space Telescope have provided one of the best measurements of the scale size of the universe to date.

The constellation
Auriga

A constellation that climbs high in the northern sky after Christmas, but is only low above the northern horizon for Southern observers. It lies along the plane of the milky way and, as a result, is where one might expect to find rich star fields and open star clusters – their stars recently formed from the dust and gas lying within the plane of our galaxy. Auriga contains three open star clusters M36, M37 and M38. The most impressive of these is M37 – an A-List object.

M37 Auriga

Open star cluster **B M**

M37, the largest, richest and brightest of the three Auriga open star clusters, was observed by Charles Messier in 1764. The cluster covers a field of view 25 arc minutes across and has a visual brightness of 6.2, so it might just be glimpsed by the unaided eye under perfect conditions. It is easily seen through binoculars, lying just to the west of the line between θ Aurigae and El Nath, (β) Tauri. It provides a lovely view with a telescope at medium power and contains over 500 stars, of which some 150 are brighter than 12.5 magnitude, and thus individually visible in a small telescope under transparent skies. One can estimate the age of an open cluster by plotting its Hertzsprung-Russell diagram (a plot of luminosity against temperature – *see* p21) and seeing at what point stars are no longer seen on

Left *This photograph depicts the main stars in Auriga, with Capella visible at the top.*

α

Capella

β

ε

η ζ

Auriga

θ

M38

M36

M37

ι

El Nath

β (Tau)

Related charts:
p107

the main sequence – the region of the plot where stars spend the majority of their lives. More massive, and hence brighter, stars leave the main sequence sooner. M37 has at least a dozen red giant stars which have evolved away from the main sequence, while the brightest stars still on the main sequence are just hotter than type F. This gives an estimated age of about 300 million years. The star cluster's distance is of order 4500 light years. Given its angular diameter this gives an overall size of the cluster of about 23 light years.

Other Messier objects in
Auriga

Sweeping north-west of M37, binoculars will pick up the two other open clusters. M36 is reached first. It is smaller than M37, about 14 light years across, and contains perhaps 60 stars. The brightest are 9th magnitude and are more massive than those seen in M37. It is thus younger, around 25 million years old, and lies at about 4000 light years away. If you continue to move in the same direction, M38 can be seen. It lies at a similar distance to M37 but, at magnitude 7.4, is less bright. At about 220 million years old, the brightest star among the 100 or so visible members of the cluster is a yellow giant of magnitude 7.9.

M36: 05h 36.3m +34° 08'

M38: 05h 28.7m +35° 51'

The constellation
Cancer

Cancer, which is a small and not very prominent constellation, lies between Gemini and Leo. It contains one A-List object – M44, the Beehive Cluster sometimes called *Praesepe*, the Latin for 'manger'. Cancer is best seen during February, March and April.

M44   Cancer

Beehive Cluster • Open cluster
E B L

M44 can be seen with the unaided eye as a misty patch just over 1° in diameter, extended slightly in a north-south direction. It lies in the triangle formed by δ, γ, and η Cancri. Overall it is of 3rd magnitude. However, the 15 brightest stars have magnitudes between 6.3 and 7.5, so those with keen eyesight should be able to resolve individual stars under perfect conditions! Binoculars or a small telescope resolve the misty patch into about 40 stars, while larger telescopes will show up to 200. Eighty are brighter than magnitude 10, and they range down to magnitude 14. It is at a distance of 577 light years and was formed about 730 million years ago. It is a little surprising that the Beehive Cluster was included in Messier's catalogue as this was designed to be a catalogue of objects that might be confused with comets. It was added just before the first Messier catalogue was published in 1771. Another star cluster, the Pleiades – M45, also not typical of the Messier objects, was added at the same time. It is suspected that these might have been included in his catalogue along with two nebulae in Orion so that it would contain more objects than the catalogue published by Lacaille in 1755!

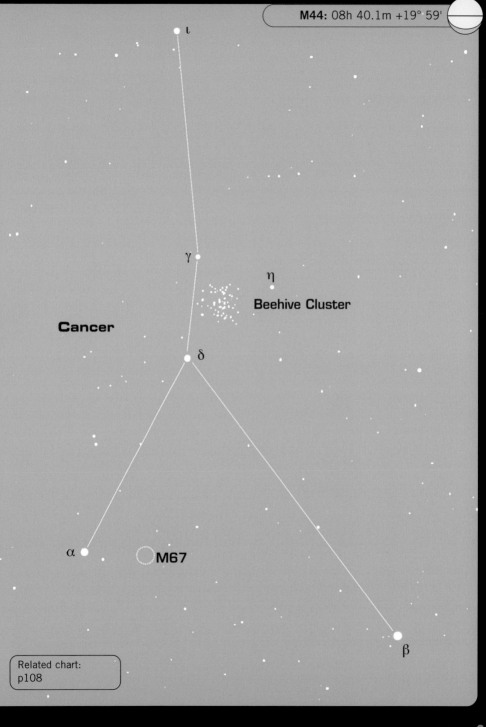

ι

γ

η

Beehive Cluster

Cancer

δ

α

M67

β

Related chart:
p108

Cancer contains a second open cluster, M67, just 2° to the west of Alpha Cancri. It is about five times further away than M44 and 6th magnitude overall. Why not look for it when you have viewed M44!

> **M67:** 08h 51.4m +11° 49'

The constellation
Canis Major

This constellation lies to the south-east of Orion and contains the brightest star in the sky, Sirius. It includes just one A-List object, the open cluster M41. This will be best seen low in the south during the winter months for Northern observers but comes almost overhead for Southern observers during their summer.

M41   Canis Major

Open cluster   E B M

M41 is easily found 4° almost exactly south of Sirius so that, for Northern observers, if Sirius is at the top of the field of a pair of binoculars or finder scope, M41 will be seen towards the bottom. (Southern observers: put Sirius at the bottom of the field and look towards the top.) With an overall magnitude of 4.5, it should be visible to the unaided eye under dark skies. It contains around 100 stars of which 50 are in the range 7th to 13th magnitude, and so should be visible in an amateur telescope. It has a beautiful orange-red star at its heart that makes a lovely colour contrast against the backdrop of fainter stars. This is a type K3 star of magnitude 6.9 and is about 700 times more luminous than our Sun.

It is thought that M41 was observed by Aristotle in 325BC, and as such would have been

Below *Canis Major is one of the most prominent constellations, and it contains the brightest star in the sky, Sirius.*

θ

γ

Sirius

α

β

M41

Canis Major

δ

η

ε

ζ

κ

Related chart:
p108

the faintest object recorded in antiquity. It was added to Messier's catalogue in 1765. M41 lies at a distance of about 2300 light years and its age is estimated at 190–200 million years.

The constellation Canes Venatici

See p170 for 'Ursa Major and Canes Venatici'.

The constellation
Carina

This southern constellation represents the keel of the ancient constellation Argo Navis – ship of the Argonauts. It contains Canopus, the second brightest star in the sky, and also the Eta Carina Nebula, in which lies the massive star Eta (η) Carina. It is best observed in March to May when Carina comes almost overhead during the evening.

C92   Carina

Eta (η) Carina Nebula • Unstable star and nebula   B M

This nebula is one of the wonders of the southern sky and the largest diffuse nebula in our galaxy. About 260 light years across, it is 7 times the size of the Orion Nebula. The Eta Carina and Orion Nebulae are called HII regions, as they contain clouds of atomic hydrogen that have been ionized by intense ultraviolet light from the many hot young O-type stars that are found within them – like those that make up the trapezium in the Orion Nebula. Eta lies in one of the brightest regions of the milky way and is best located by starting at the constellation Crux and moving

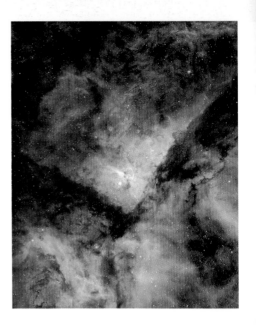

along the milky way westwards 13° – about two and a half binocular fields. On the northern edge of the nebulosity lies the star Eta (η) Carina. It is about 100 times the mass of the Sun and many times its luminosity. A highly unstable star, it is ejecting material into the surrounding space making it highly variable in brightness. In 1843 it became the second brightest star in the sky, but then slowly faded from view, hidden by the ejected shroud of gas, reaching a minimum magnitude of 7.6 in 1968. Its brightness has since been rising and is now about 4.2 magnitudes – so visible to the unaided eye again. The outburst in 1847 was the biggest explosion that any star has been known to survive and it resulted in the ejection of two lobes of expanding gas, now about 0.8 light years apart, which have been beautifully imaged by the Hubble Space Telescope.

Above *Carina is an impressive southern constellation, which contains the second brightest star in the sky, Canopus.*

u

η

Eta Carina Nebula

p

Carina

θ

**Southern
Pleiades**

Related chart:
p113

1318

The constellation
Centaurus

Though not as well known as Crux, there is no doubt that Centaurus is the most impressive southern constellation. The ninth largest constellation in the sky, over 100 of its stars are visible to the unaided eye. Centaurus lies too far south to contain any Messier objects. It does, however, contain three A-List objects: one of these, Alpha (α) Centauri, is the nearest star system to our Sun, another is the best globular cluster in the heavens and the third is a dynamic galaxy. Centaurus is best observed in the autumn when it comes almost overhead.

Centaurus

Alpha (α) Centauri or Rigil Kentaurus • Multiple star system
E H

To the eye, Rigil Kent looks like a single star of magnitude -0.3, the third brightest star in the sky. It lies at a distance of 4.35 light years. With a telescope it splits easily into two stars: the primary A is a G-type star very similar to our Sun which has an apparent magnitude of -0.04, while the secondary B is an orange K-type star of magnitude 1.2. They circle each other in a highly elliptical orbit with a period of 80 years, having a mean separation of 24 AU (1 AU is the mean Earth-Sun distance – so 24 AU is about the distance of Uranus from the Sun). They currently appear separated by 19 arc seconds and are thus easily split with a small telescope. There is a third member of the star system, Alpha (α) Centauri C, which is 13,000 AU from A and B. As it is measurably closer to us, at a distance of 4.22 light years, so

it is also called Proxima Centauri as it is the nearest star to our solar system.

C80   Centaurus

Omega (ω) Centauri
• Globular cluster   E B M

This is the most spectacular globular cluster in the heavens, appearing as a 4th magnitude 'fuzzy' star to the unaided eye. Moving north of Hadar, Beta (β) Centauri, by 5° and 10° respectively, about two finder fields in total, are two 2nd magnitude stars Epsilon (ε), the more southerly, and Zeta (ζ) Centauri. Omega (ω) Centauri makes an almost equilateral triangle with these two stars on its eastern side. With an equatorially mounted telescope one can centre Zeta (ζ) Centauri in a medium-power telescope field, lock the declination axis and then sweep 28 minutes (of time -7° in angle) west in RA. This should bring Omega Centauri into the field of view. It contains perhaps 10 million stars in a region about 160 light years across. It lies at a distance of some 16,000 light years and has an apparent diameter of 30 arc minutes or more – the size of the Full Moon! It does not have a very bright central core as is found in 47 Tucanae. Omega Centauri is a wonderful sight in a telescope of 8 inches (200mm) or more, but still very rewarding even with a 4in (100mm) aperture if the sky is dark and transparent. With a mass of about 5 million solar masses it is 10 times more massive than the Milky Way's other large globular clusters such as M13 in Hercules and has a similar mass to some small galaxies! It is the most luminous globular cluster in our Milky Way and in our Local Group of galaxies is only outshone by the cluster named G1 in the Andromeda Galaxy.

C80: 13h 26.8m -47° 29'

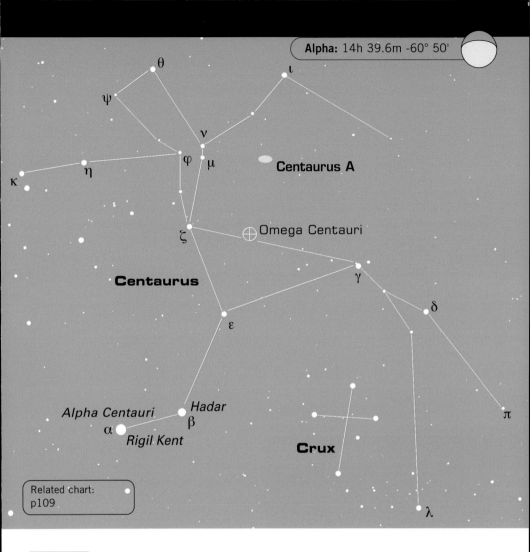

θ

ψ

ι

ν

φ μ

Centaurus A

κ

η

ζ

⊕ Omega Centauri

Centaurus

γ

δ

ε

Alpha Centauri   Hadar

α   β

π

Rigil Kent

Crux

Related chart:
p109

λ

C77 — Centaurus A

NGC 5128 • Active galaxy

E B M

Centaurus A, so called because it is a very bright source of radio emission, is a large bright (7th magnitude) elliptical galaxy crossed by a very prominent dust lane. It is often rightly called a 'peculiar' galaxy and is one of the most interesting galaxies that we can observe. One possible scenario is that it may have 'eaten' a large spiral galaxy in the last few billion years. The distance to C77 is not well known but probably lies in the range 10–16 million light years. It forms a right-angled triangle with Zeta (ζ) Centauri and C80 Omega, (ω) Centauri, (see instructions for finding both in the C80 description). Using an equatorial telescope find three 3rd magnitude stars, μ, ν and φ Centauri, forming a tight triangle 5° north of Zeta (ζ) Centauri. With μ in the centre of a wide field of view, lock the declination axis and move westwards 24 minutes in right ascension (6°) to bring C77 into the field of view. The galaxy subtends an area 17 by 13 arc minutes in size. With a small telescope and dark skies, both halves of the galaxy are visible with the dark dust lane about 40 arc seconds in width running across the middle. Larger telescopes and higher power will show two foreground stars superimposed on the southern half, the brighter being 12 and the fainter 13.5 magnitude. Centaurus A is one of the largest, most massive and luminous galaxies known. It harbours an 'active galactic nucleus' in which material is falling in towards a supermassive black hole – perhaps weighing 100 million times that of our Sun! Opposing jets of particles have left the active nucleus and radiation from them, producing two vast lobes of radio emission above and below the galaxy. These make Centaurus A one of the strongest radio sources in the sky.

C77: 13h 25.5m -43° 01'

Though Crux is perhaps the best known of the southern constellations, it is not as prominent as one might expect. Lying right along the plane of the milky way it tends to recede into the great mass of stars under dark skies. It is, however, easily found by following the line from the very bright stars Alpha (α) through Beta (β) Centauri which leads directly to Crux. It includes three A-List objects. However, as one, the Coal Sack, is actually a lack of anything to see, should it really be called an object?

Crux

Alpha (α) Crucis – Acrux
• Double star   E H

Alpha Crucis was too far south to have been given an ancient name, so Acrux is simply a combination of the A in alpha and Crux. Being of 1st magnitude (0.83), it is the 12th brightest star in the sky. Under high power, a telescope reveals it to be a binary system with two very similar B-type stars having magnitudes of 1.33 and 1.73, which are separated by 4 arc seconds. With surface temperatures of around 27,000K, they are highly luminous. In fact, the brighter star is itself a double, the two component stars orbiting each other every 76 days are too close to split with a telescope. Therefore, Acrux is a triple star system.

Left *The four brightest stars of the southern constellation Crux form a cross, which is why it is also referred to as the 'Southern Cross'.*

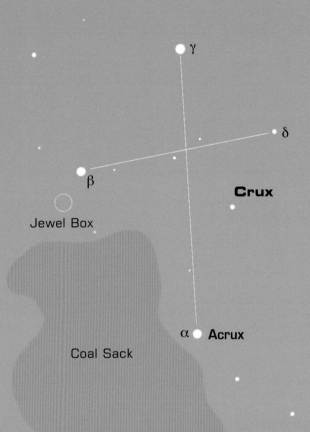

γ

δ

β

Crux

Jewel Box

α **Acrux**

Coal Sack

Related chart:
p113

C94 — Crux

The Jewel Box • Open cluster
B L

This is an open cluster, also called Kappa Crucis, which contains about 100 visible stars and is about 10 million years old. It lies some 7500 light years away and spans a 10 arc minute field of view, so filling a volume of space about 20 light years across. Lying close to Beta (β) Crucis, it is easy to find and is best seen with binoculars or a telescope at low power. It contains many highly luminous blue-white stars along with a central red supergiant that makes a beautiful colour contrast. It was named the Jewel Box by Sir John Herschel who called it 'a gorgeous piece of fancy jewellery'.

C94: 12h 53.6m -60° 21'

C99 — Crux

The Coal Sack • Dark nebula
E B

Just to the south of the Jewel Box is a pear-shaped region of obscuring, or dark, nebula, 7° long by 5° wide. Called the Coal Sack, it is a dense region of dust and gas about 2000 light years from us that is hiding the light from more distant stars. It is the most prominent and conspicuous dark nebula along the plane of the Milky Way, and is easily picked out by eye as a big dark region against the bright band of light from the stars making up our galaxy. It will fill the field of view of all but the widest field binoculars.

C99: 12h 52m -63° 18'

The constellation
Cygnus

The northern constellation Cygnus lies along the plane of the Milky Way and its brightest stars can get lost against the magnificent backdrop of stars if the sky is too dark and clear! These stars form what is called the Northern Cross. Cygnus contains over 11 open star clusters but none have made the A-List. Just two objects have: a beautiful double star system and the faint nebulosity of a fading supernova explosion. Cygnus is best observed during the months from August to November.

Cygnus

Albireo – Beta (β) Cygni • Double star
E B H

This is perhaps the most beautiful double star in the sky with a wonderful colour contrast between the brighter component, at magnitude 3 and yellow-gold or amber in colour and its fainter, 5.1 magnitude, companion which is a vivid blue-green. They are separated by 34 arc seconds, so any telescope will split them even under the worst of seeing conditions. If 10x50 binoculars are held very steadily or mounted on a tripod, they should also be able to show that Albireo is a double. (This is where image stabilized binoculars would be very useful!) They are 380 light years away and the primary component is a K-type star that has evolved off the main sequence, while its fainter companion is a hotter but smaller, B-type, main sequence star. This pair provides us with observable proof that stars evolve. As the two stars lie at the same distance, we know that the yellow-gold star is six times more luminous than the blue star. Stars lying on

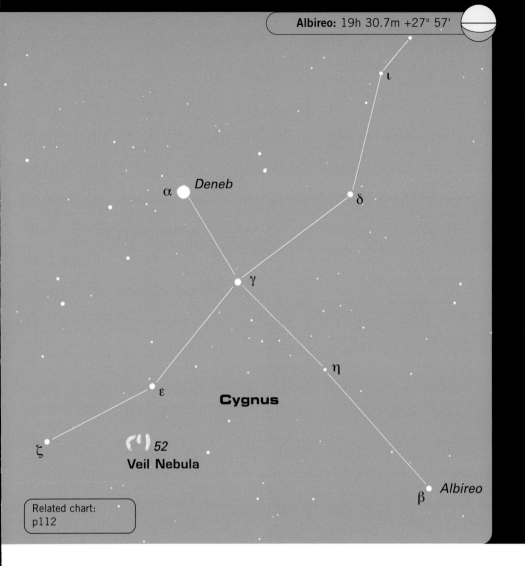

Deneb

α

δ

ι

γ

η

Cygnus

ε

52

Veil Nebula

ζ

β   Albireo

Related chart:
p112

the main sequence burn hydrogen into helium in their cores. The more massive stars are brighter and hotter, emitting blue or white light, while the less massive ones are cooler and less bright, emitting yellow, orange or red light. A main sequence yellow star would therefore be far less bright than a blue star. So the yellow star in Albireo cannot be a main sequence star.

In fact, it has evolved away from the main sequence and has become a yellow giant as it converts hydrogen and helium into heavier elements in its core. It has become far bigger in size and, although each square metre of the surface emits less light than that of the blue star, its surface area is so much larger that overall it emits six times more light.

The Veil Nebula
• Supernova remnant   B L

This is one of the most challenging of all the A-List objects to observe, as it requires very dark and transparent skies to make out the faint nebulosity against the sky background. If you can find an observing site with little or no light pollution and the skies are transparent – perhaps when heavy rain has washed the dust out of the atmosphere – then parts of the Veil Nebula can be seen even with 8x40 binoculars. The Veil Nebula is the wispy nebulous remnant of a star that exploded some 5000 years ago in what is called a Type II supernova. For a brief period its brightness could have rivalled that of the Moon and it would have been visible in broad daylight. There are three main areas of nebulosity in a roughly circular outline: NGC 6992 and 6995, making up Caldwell 33, to the east with NGC 6960, C34, to the west. C34, though somewhat less bright than C33, is easier to find as it runs north-south 'through' the 4.2 magnitude star 52 Cygni. This star is 3° south of Epsilon (ε) Cygni, the eastern star of the Northern Cross. So centre this in your binoculars or finder and drop south by less than one field width to find 52 Cygni and, hopefully, the faint hints of nebulosity. The slightly brighter parts of the nebulosity that make up C33 are 2.5° to the east and very slightly north so the whole of the veil can be easily encompassed using 8x40 or 10x50 binoculars. Using a telescope, only a 2-inch, wide-field, eyepiece and a short focal length could encompass the whole field. In any event, use your lowest power eyepiece. Using an equatorially mounted telescope, find 52 Cygni first and try to observe the nebulosity there, then, with the declination axis locked, sweep eastwards by 2.5°. Lock the right ascension axis and inch up, perhaps half a degree in declination to observe the nebulosity of C33. At the same declination as C33 moving two-thirds of the way back towards C34, you may glimpse a third area of nebulosity running north-south through a slightly inclined chain of stars. This is one object where the use of an ultra-high-contrast or OIII filter could really help. Good hunting!

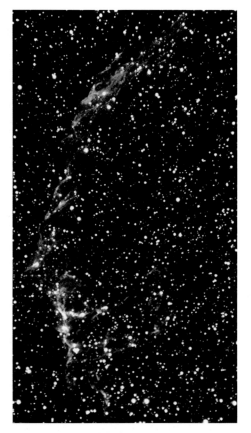

C33: 20h 56.0m +31° 43'
C34: 20h 45.7m +30° 43'

Left *The Veil Nebula is a remnant of an exploded star. It lies in the southern part of the constellation Cygnus.*

α *Deneb*

δ

ι

α

γ

β

δ

η

Cygnus

ζ

52
Veil Nebula

β *Albireo*

Related chart:
p112

Dorado

This constellation, close to the south celestial pole, is highest in the sky during the winter months. It lies south and a little east of the very bright southern star Canopus. Within its boundaries lies the Large Magellanic Cloud (LMC), an A-List object, while another, 30 Dorado, lies within the LMC itself.

cloud seen against a dark sky, or a part of the milky way that has somehow become detached. It must have been known since humans first looked up at the southern sky, but was 'discovered' by the Portuguese explorer Ferdinand Magellan in 1519. The LMC has an angular diameter of 6° and will thus nicely fill the field of view of most binoculars. A telescope used at low power will be able to sweep across the galaxy picking up the many bright nebulae and open star clusters. The most spectacular of these merits an A-List entry in its own right.

LMC   Dorado

LMC – Large Magellanic Cloud
• Irregular galaxy   E B L

The LMC is an irregular (or possibly barred spiral) galaxy that, at about 170–180,000 light years' distance, is the second closest galaxy to our own Milky Way (a dwarf elliptical in Sagittarius is closer). It is at least 50,000 light years in diameter and contains several billion stars. The LMC is the fourth largest galaxy in our local group of galaxies (dominated by the Milky Way, M31 and M33). To the unaided eye it appears just like a

C103   Dorado

30 Doradus • Bright Nebula
and open cluster   B M

The name 30 Doradus is used collectively for a bright nebula, commonly called the Tarantula Nebula due to its similarity in appearance to the

Below *In the star-forming region 30 Doradus Nebula, in the LMC, new stars are born within collapsing gas and dust clouds. At the nebula's centre is a spectacular cluster of massive stars.*

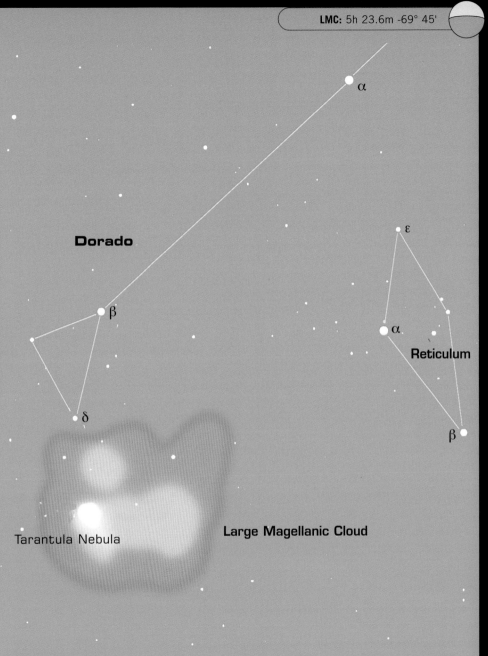

α

ε

Dorado

β

α

Reticulum

δ

β

Tarantula Nebula

Large Magellanic Cloud

Related chart:
p113

spider of that name, and the cluster of stars embedded within it. The name for this object comes from the fact that together they were first catalogued as a star. The Tarantula Nebula is an immense star-forming region, vastly bigger than the Orion Nebula, where clouds of gas are being excited by the ultraviolet radiation emitted by the very hot young stars that have formed within it. In fact, were it as close as the Orion Nebula, it would cover an area of sky 30° across! It is the most massive HII region, as these regions consisting mainly of ionized hydrogen are called, in the entire local group of galaxies. Lying at a distance of about 165–170,000 light years, it is over 3000 light years across. (For comparison, the Orion Nebula is 40 light years across.) It contains a vast number of blue supergiants, type-O stars that are among the most massive and luminous known – around 100 times the mass of the Sun and perhaps 100,000 times as luminous. Blue supergiant stars have short lives and evolve quickly ending their lives in spectacular supernova explosions. Astronomers were treated to just such an event in 1987; the resulting remnant, 1987A, has been studied extensively ever since. To the unaided eye the 30 Doradus region, about the angular size of the Full Moon, appears rather like the Lagoon Nebula, M8, in Sagittarius (*see* p161). Visually about 4th magnitude, it is best seen with binoculars and telescopes. The brightest stars within the cluster are between 14th and 12th magnitude, and so can be seen individually under dark skies and good seeing. Even the smallest telescopes will reveal the complex structure, showing loops of excited gas and embedded stars. With telescopes of eight or more inches aperture (>200mm) the visual appearance can even approach that seen in the wonderful photographs of the region.

C103: 5h 38.6m -69° 05'

Gemini

Castor – Alpha (α) Geminorum
• Multiple star system

E H

This is a visual double star made up of two blue-white stars, A and B, of 1.9 and 2.9 magnitudes respectively. The pair orbit each other every 400 years and are now as close as they ever get, making them somewhat of a challenge to split and requiring very good seeing. In fact, their spectra show that each is itself a double star! Castor A is made up of two identical 2 solar mass stars orbiting each other every 9.2 days, while the stars that make up Castor B orbit even faster, every 2.9 days. One minute of arc to the south will be seen a faint 9th magnitude star. This is also part of the Castor system and is itself a double star consisting of two M-type dwarf stars about 0.6 solar masses. Amazingly, they are only twice the diameter of our Sun apart, and orbit each other every 2 hours. So Castor is actually a sextuple star system that would look absolutely amazing should one be able to pass close enough on a space journey!

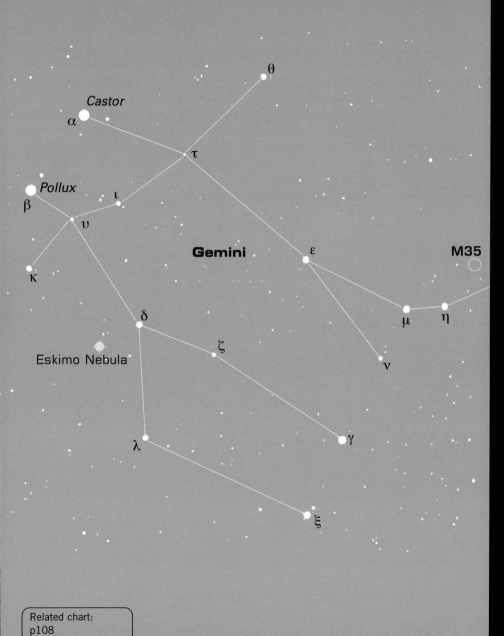

θ

Castor
α

Pollux
β   ι
   υ

κ

Gemini   ε   **M35**

μ   η

δ

ζ

Eskimo Nebula   ν

λ   γ

ξ

Related chart:
p108

Open cluster E B L

Just north of the left foot of the northern-most twin, the unaided eye can see a hazy patch of light whose size is somewhat more than that of the Full Moon. Binoculars will resolve this into individual stars spread uniformly within the cluster. The brightest are 8th to 9th magnitude. It is probably best seen in a telescope at low power using a wide-field eyepiece. M35, also known as NGC 2168, is thought to contain about 500 stars within a volume about 24 light years across and lies at a distance of 2700 light years. Given a very dark site and good sky transparency you may also observe a small compact cluster, just 5 arc minutes across, that lies half a degree south-west of M35. This is NGC 2158, a cluster that, overall, has a visual magnitude of about 8 to 9. It is actually very similar in size to M35 but some six times more distant so appearing both smaller and fainter.

M35: 06h 08.9m +24° 20'

The Eskimo or Clown Nebula
• Planetary nebula H

This is a planetary nebula that has an unusually bright 10th magnitude central star that is the white dwarf remnant of the stellar explosion that gave rise to the nebula. The nebula got its name from the fact that the central bright region forms a face (with the white dwarf as its nose) while a larger, more diffuse, outer ring makes up the fur hood of the Eskimo's parka or the ruff of the clown's outfit. Sadly, with a small telescope, this outer ring will not be visible except under very dark skies, but you will easily see the central

region and white dwarf. An 8in (200 mm) telescope will give you a reasonable chance of detecting the outer envelope even in less good conditions. It lies quite close (2° 21') to the 3.5 magnitude star Delta (δ) Geminorum. To the south and east of δ Geminorum lie three 5th magnitude stars: 56, 61 and the double star, 63 Geminorum forming a right angle triangle. The Eskimo nebula lies just 37 minutes of arc to the south and a little east of the left hand star of the double. With a low-power eyepiece it will appear as a 'star' just to the south of an 8th magnitude star just 1.6 arc minutes away – the pair forming an apparent double. Increasing the magnification will reveal the true nature of the nebula as a 10th magnitude central star surrounded by a bright region of nebulosity. Its visual magnitude is about 9th magnitude. Incidentally, δ Geminorum lies just 10 arc minutes from the plane of the ecliptic (the path of the Sun across the sky) and it was very close to both this star and the Eskimo Nebula that Clyde Tombaugh discovered the planet Pluto in 1930. As these notes were being written, the point was reinforced as another planet, Saturn, passed just 4 arc minutes away from the star!

C39: 07h 29.2m +20° 55'

In May, June and July the constellation Hercules comes almost overhead for those in the northern hemisphere. It lies roughly half way between the two bright stars Vega, in Lyra, and Arcturus, in Bootes. During those same months it can be seen just above the northern horizon from the southern hemisphere, but its low elevation may make the two A-List objects, M13

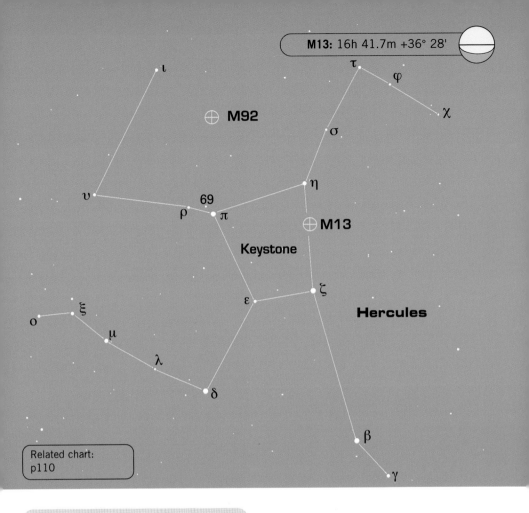

M13: 16h 41.7m +36° 28'

ι

τ

φ

χ

⊕ M92

σ

η

υ

69

ρ π

⊕ M13

Keystone

ξ

ε

ζ

o

Hercules

μ

λ

δ

β

γ

Related chart:
p110

and particularly M92, difficult to observe from the southern hemisphere. (At Sydney, M92 rises 12° above the horizon, M13 rises 20° above.)

M13   Hercules

Hercules Cluster
• Globular cluster   E B M

This is arguably the finest globular cluster in the northern sky and was discovered by Edmund

Halley in 1714. Charles Messier logged it as he tracked the path of a comet in 1779. However, with a visual magnitude of 5.8 (the noted observer, James O'Meara puts it at 5.3) it is visible to the unaided eye so may well have been seen in antiquity. M13 lies along the western side of the 'keystone' about 2.5° south of the star Eta (η) Herculis – it is thus very easy to find. M13 contains several hundred thousand stars in a volume of space 145 light years across. It has an angular diameter of 20 arc minutes and lies around 25,000 light years from us. These statistics hide the sheer beauty of the cluster that needs a really dark and transparent sky and good seeing to observe at its best. (These do not often occur together!) The dark sky enables fainter stars to be seen and, if the atmosphere is steady, the stellar images will be 'tighter' allowing even fainter stars to stand out against the background glow of light from the cluster. It will look good in any telescope, however small, but the view through a 10in (250mm) or larger telescope under perfect conditions can take your breath away, with M13 looking almost three-dimensional in appearance. Curving arcs of stars appear to extend southeast and northwest, branching out into the surrounding space. Several stars reach 11th and 12th magnitudes with 20 or more at 13th magnitude.

M92   Hercules

Globular cluster   B M

M92 is also a superb globular cluster, somewhat overshadowed by its near neighbour M13. With a visual magnitude of 6.5, it is just on the limit of unaided eye visibility but really needs binoculars or a telescope to observe. M92 is north and a little east of the north-eastern star of the keystone's, 3rd magnitude Pi (π) Herculis. With an equatorially mounted telescope, centre the finder or low-power field of view on this star and then move very slightly northeast to the 5th magnitude star 69 Herculis. Lock the RA axis then sweep north in declination just under 6°. M92 will then be in the centre of the field. It lies at a distance of about 27,000 light years and has an angular extent of 14 arc minutes corresponding to a diameter of 109 light years. The total mass of the stars is of order 300,000 solar masses. At a very similar distance from us as M13, some of the brightest stars within it will be individually resolved with good seeing and transparent skies. Remember that fainter stars can be

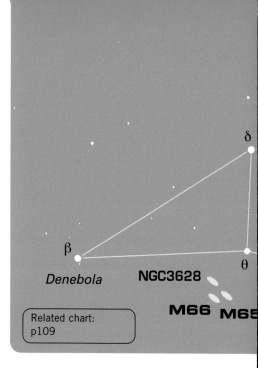

Related chart:
p109

seen when 'averted vision' is used – just look slightly away from the centre of the cluster.

M92: 17h 17.1m +43° 08'

The constellation
Leo

Leo follows Orion across the sky so as Orion is setting, Leo is highest in the sky. It is seen best in March, April and May. Leo lies away from the obscuring dust of the milky way. This means that it is devoid of any open clusters or star-forming regions but, on the other hand, the lack of obscuring dust allows distant galaxies to be seen. Two pairs of Messier galaxies are included in the A-List.

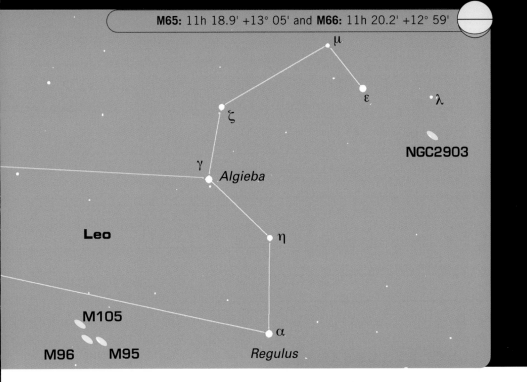

μ

ε

λ

ζ

NGC2903

γ

Algieba

Leo

η

M105

M96   **M95**

α

Regulus

M65 • M66

Spiral galaxies

B L M

A pair of 9th magnitude spiral galaxies visible either together in a low-power eyepiece or individually at medium power. M65 is a type Sa spiral that lies at a distance of 35,000,000 light years and has a magnitude of 9.3. M66, considerably larger than M65, is a type Sb spiral lying slightly further away at 41,000,000 light years and fractionally brighter at magnitude 8.9. These two galaxies are located halfway between Theta (θ) and Iota (ι) Leonis, and just to the east

Right *M65 is a type Sa spiral galaxy with a prominent nucleus and tightly wound spiral arms.*

of the 5th magnitude star 73 Leonis. They can be seen, along with a third galaxy, NGC 3628, just to the north, with a pair of 8x40 or 10x50 binoculars – providing that the sky is very dark and transparent. Sadly, in our light-polluted skies such conditions do not occur often. The 9th magnitude brightness given for these galaxies seems quite bright, but this is the integrated magnitude over the whole galaxy and each part only shines with the equivalent brightness of a 12th magnitude star. These 'faint fuzzies', as they are often called, are quite a challenge!

M95 • M96   Leo

Spiral galaxies   L M

This pair of galaxies, separated by just 42 arc minutes, are located close to another Messier galaxy, M105. They lie between the 5th magnitude stars 52 and 53 Leonis and are almost due east of Regulus. A good way to find them with an equatorially mounted telescope is to put Regulus towards the southern side of the finder scope or low-power telescope field of view, lock the declination axis and move just under 9° east. This should bring you to M95, with M96 a further 42 arc minutes east of M95. M95 is a barred spiral of Hubble type SBb lying at a distance of 38,000,000 light years with an integrated visual magnitude of 9.7. Under ideal conditions it looks a little like Saturn; a central concentration of light – the nucleus – looking like the planet and the bar looking like the rings. This is visible with a 4in (100mm) telescope but a larger aperture will certainly help! M96 is slightly further away, at 41,000,000 light years, and is a type Sa galaxy with a magnitude of 9.2. It has a very condensed central core and can take on the appearance of an eye. These objects are not the

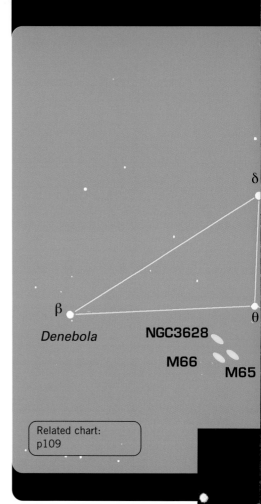

easiest to see but, when you do succeed, reflect on the fact that you are looking back in time many millions of years!

Right M95 is a type SBb barred spiral having nearly circular spiral arms attached to the bar across the central nucleus.

μ

λ

ε

ζ

γ *Algieba*

Leo

η

M105

M96 **M95**

α

Regulus

This contains two A-List objects: a wonderful multiple star system and a planetary nebula that is easy to find and observe. These are best observed in July, August and September, but are sadly rather low above the northern horizon for observers in the southern hemisphere.

Lyra

Epsilon (ε) Lyrae – The Double double • Multiple star system
E B H

This is perhaps the best multiple star system that can be easily observed! Look first with binoculars and find Vega, at magnitude -0.04

Below *M31 – the Andromeda Nebula – is our nearest large galaxy just under three million light years distant. It is similar to our Milky Way.*

the fifth brightest star in the heavens and one of the stars that make up the Summer Triangle. Centre the field of view on Vega and, to the east and a little north, the two 'stars', epsilon[1] and epsilon[2], that make up Epsilon (ε) Lyrae will be seen as a double, separated by 208 arc seconds. Once found you may, given keen eyesight, be able to 'split' the double using just your eyes. Each of the two 'stars' is just brighter than 5th magnitude.

If you observe them with a telescope using reasonably high power when Lyra is high in the sky and the atmosphere is steady, you should see that each of the two 'stars' is itself a double, one orientated along the line joining the two major components, the other at right-angles to it. The former pair (epsilon[1]) are of magnitude 4.6 and 5.0 separated by 2.6 arc seconds with the latter (epsilon[2]) having magnitudes of 5.1 and 5.5 and separated by 2.3 arc seconds – a small separation, which is why the seeing has to be good to split them. They all appear white to the eye. Each pair is orbiting each other with periods of about 1200 (epsilon[1]) and about 600 (epsilon[2]) years, while the two pairs are also gravitationally bound and orbit each other about once every million years.

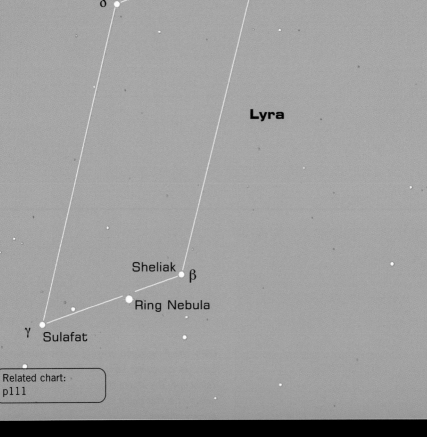

Epsilon Lyrae

ε¹ ε²

α
Vega

ζ

δ

Lyra

Sheliak β

Ring Nebula

γ Sulafat

Related chart:
p111

M57   Lyra

The Ring Nebula
• Planetary nebula   H

Perhaps the easiest planetary nebula to observe, the Ring Nebula looks like a smoke ring or doughnut. It lies just below the line joining Gamma (γ) and Beta (β) Lyrae, being slightly closer to β Lyrae. Just sweep a telescope between the two stars and, given reasonably dark skies, it should be immediately obvious. It has a magnitude of 8.8 and is 1.4 by 1.0 arc minutes in angular size. It is now believed that it is a shell or possibly a cylinder of bright glowing gas that was blown off when the progenitor star exploded leaving the very hot (~100,000K) white dwarf star that is seen at its centre in photographs. This bluish compact object, about the size of the Earth, is only 15th magnitude and so, while showing up well in photographs, will only be seen in the biggest backyard telescopes. Ultraviolet light emitted by the white dwarf excites the surrounding gas to glow. The most recent estimates give M57 an age of about 7000 years. It is thought to lie at a distance of just over 2000 light years and is about half a light year across – or 500 times the size of the solar system.

M57: 18h 53.6m +33° 02'

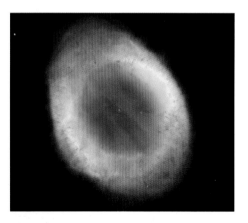

The constellation
Orion

Probably the best known of all the constellations, Orion dominates the sky in January, February and March. Perhaps surprisingly, given its closeness to the milky way, it is rather devoid of deep-sky objects that merit inclusion the A-List. However, the one that does is one of the true highlights in the heavens.

M42   Orion

Orion Nebula • Bright nebula
E B L M H

The Orion nebula, seen as a diffuse glow in the sword of Orion, is one of the most beautiful objects in the heavens. It is a region of star formation 1600 light years from us, the glowing gas excited by ultraviolet light from the young, very hot stars at its heart. It extends over a region of angular size 1 by 1.5° but is part of a much larger cloud that covers much of the constellation, over 10° across. Photographs show swathes of nebulosity such as 'Barnards Loop' arching around the M42 region. Seeing that M42 is one of the first deep-sky objects that most astronomers observe – and rightly so – it is surprising that there is no mention of it in ancient records. Apparently, not even Galileo observed it, even though, at about 4th magnitude, it is visible to the unaided eye under reasonably dark skies. The gases that make up the

Left *The Ring Nebula, lying in the direction of Lyra, is captured here by the Hubble Space Telescope; the image shows the expanding outer layers of a star being shed during its late evolutionary phases.*

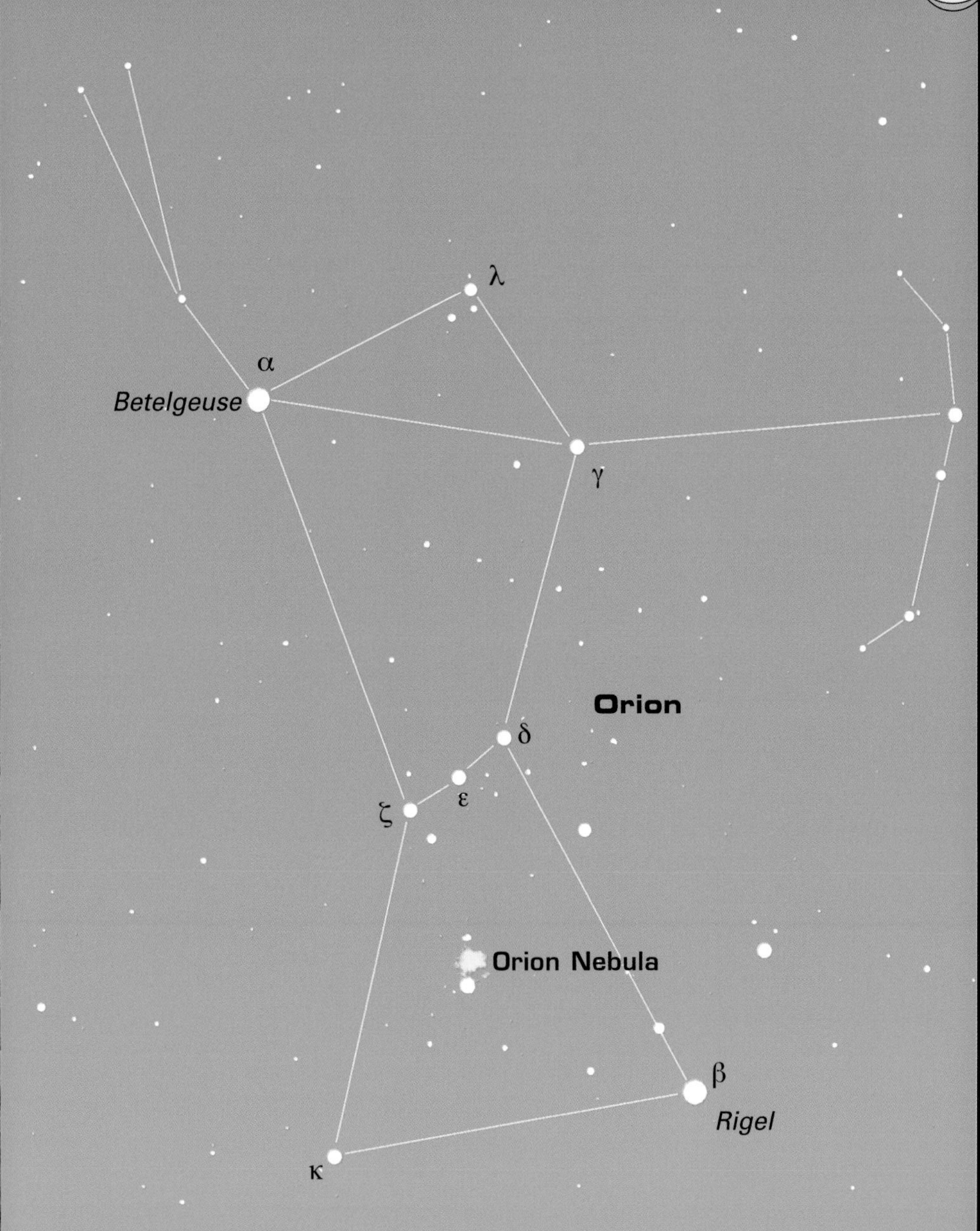

λ

α
Betelgeuse

γ

Orion

δ

ε

ζ

Orion Nebula

β
Rigel

κ

Related chart:
p107

nebula are predominantly hydrogen and helium dating from the origin of the universe along with nitrogen and oxygen produced by nuclear fusion as stars evolve. The electrons are stripped from their nuclei by the ultraviolet light and, as they re-combine, give off well-defined spectral colours: a pinky red for hydrogen and green and blue for oxygen. The eye is not sensitive to red so with a reasonably large telescope, the brighter parts of the nebulosity appear greenish in colour. Observing the nebula under dark and transparent skies with a telescope at low power shows the wonderful looping whirls of dust and gas seen across a wide region. Using a medium-power eyepiece, the central region of the nebula will be seen to harbour a number of bright stars; three almost in a line and 4 making up what is called the 'trapezium'. A dark dust cloud can be seen intruding in towards the bright inner core of the nebula surrounding the trapezium. This is called the 'fish's mouth'. Moving to high power will show the four stars, A to D, of the trapezium. The brightest is 5th magnitude and provides the vast majority of the ultraviolet light that excites the gas in the nebula.

Two others are 6th magnitude with the faintest 8th magnitude. Using very high magnification under nights that are both dark and have good seeing (which are rather rare), even a 4in telescope will pick out a fifth, 10th magnitude, component called E, making up a flattened triangle with the closest pair of stars. It lies just outside the trapezium. There is a sixth, 10th to 11th magnitude, star called F on the opposite side of the trapezium but this is much harder to see.

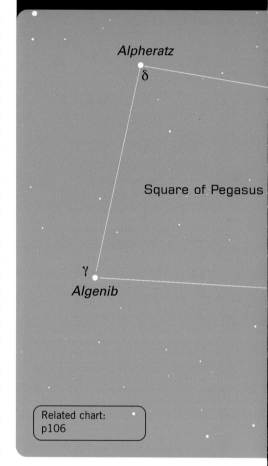

Related chart:
p106

The constellation
Pegasus

Pegasus is at its highest in the sky during September, October and November and its square – the body of the winged horse – is very well known. A very good test of sky transparency is to count the number of stars that you can see with your unaided eye within the square. If you count four, then the transparency is reasonably good, but if you can see more than four, then it will be an excellent night for hunting down faint galaxies and nebulae! Pegasus is away from the

η

β
Scheat

μ
λ

ι
κ

51 Peg

Pegasus

Markab
α

M15
⊕

ζ
ε
Enif

θ
Baham

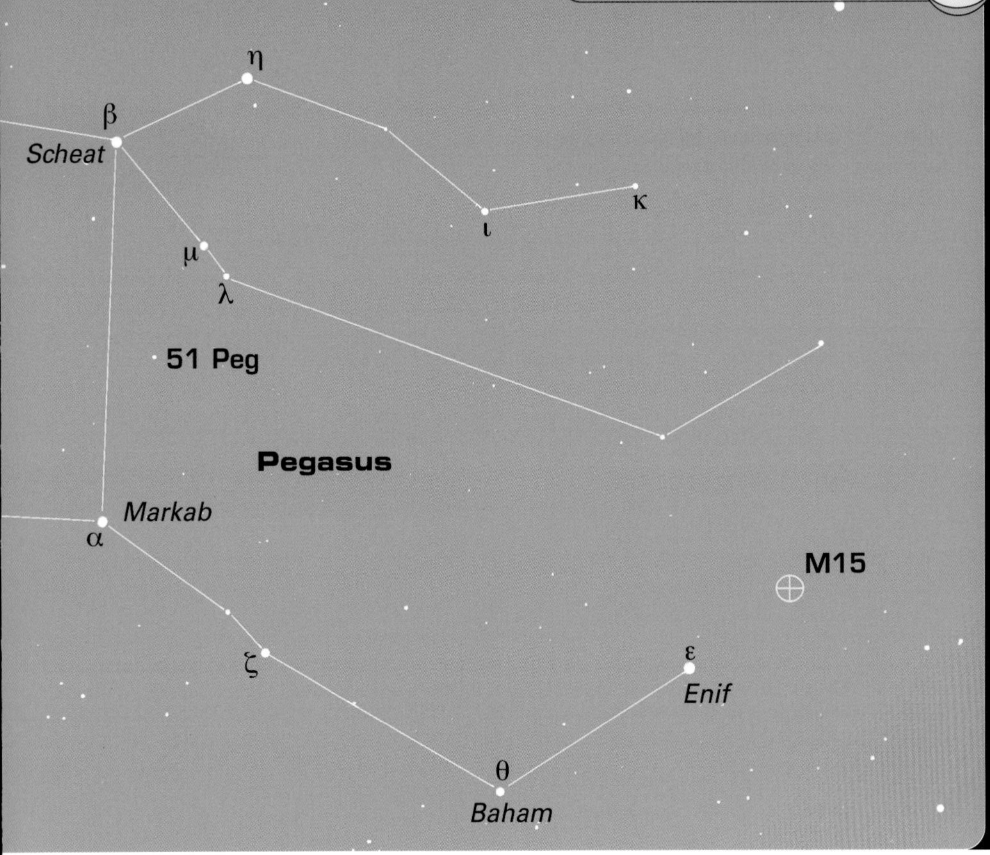

plane of the Milky Way so is not rich in deep-sky objects but there are two A-List objects within its boundaries.

M15 · Pegasus

Globular cluster · B M

M15 is one of the six globular clusters brighter than 7th magnitude in the northern sky. It lies some 30,000 light years away beyond the thick dust lanes and star clouds in Sagittarius. M15 is easily found by following the line of the two stars Theta (θ) and Epsilon (ε) Pegasi a further 4° in a north-westerly direction. So, by placing Epsilon at the appropriate side of a binocular or finder scope field, M15 should be visible towards the other. With medium power, the cluster is seen to lie within a triangle of three stars, one 7th and two 8th magnitude. M15 lies at a distance of 37,000 light years. Visually it appears to have an angular extent of about 7 arc minutes and has a very bright and compact core. Its overall brightness of magnitude 6.2 means that, under ideal viewing conditions, it could be seen with the unaided eye. Its brightest individual stars are 12th to 13th magnitude, so may be picked out individually against the glow of the fainter stars in the cluster when observing with a large aperture telescope. A recent Hubble Space Telescope image shows that stars in the dense core of M15 are crowded together closer than anywhere else in our galaxy, except at its very heart. At least 30,000 stars pack into a volume just 22 light years across. This could be the result of a process called 'core collapse' or perhaps a massive black hole is hidden at its centre!

51   Pegasus

51 Pegasi • Star with planet   B

Perhaps this is an odd choice for the A-List. There is nothing to see except a single 5.49 magnitude star that can be found just to the west (about 2°) of the line joining the two stars at the western side of the square, Alpha (α) and Beta (β) Pegasi. It is a little nearer to Alpha. So why is it included? Simply because it has a place in history – the first Sun-like star around which a planet was detected. The planet, called 51 Pegasi b, has a mass of at least 0.47 times that of Jupiter. It orbits 51 Peg every 4.23 days; compare that with the 88 days that it takes Mercury to orbit the Sun! Its circular orbit has a radius of only 0.05 AU, just 7.5 million kilometres (4.6 million miles). So close to its star, its atmosphere would be extremely hot, perhaps 1000°C, and it would suffer extreme tidal forces. No one expected that a giant planet could exist so close to its sun – giving theoreticians much food for thought.

So how was it discovered? The light reflected by it would be lost in the glare of its sun so it cannot be seen. However, both the planet and star orbit their common centre of gravity so 51 Peg is moving round in a small circle. This means that sometimes it is coming towards us and sometimes away from us giving a very small shift in its spectral lines due to the Doppler shift. It

was this that was detected in the spectrum of 51 Peg so allowing the period and minimum mass of its planetary companion to be calculated. Since then, many other planets have been detected by this method, but 51 Pegasi b will always have the distinction of being the first!

51 Pegasi: 22h 57.6m +20° 46'

The constellation
Perseus

Perseus lies along the plane of the Milky Way, but as we view it, we are looking away from the galactic centre, so that there are fewer clusters within its boundaries than when looking towards Sagattarius or Scorpius. However, the two clusters that form the Perseus Double Cluster provide one of the best binocular sights in the heavens and Perseus also harbours a very interesting star system, Algol. Whilst, for Northern observers, Perseus comes high overhead in the latter months of the year, for Southern observers it only just rises above the northern horizon so that, whilst they should be able to observe Algol, sadly, the Double Cluster will lie below their horizon.

C14   Perseus

The Double cluster • Twin open clusters   E B L M

Visible to the unaided eye as a hazy patch in the Milky Way, binoculars or a small telescope at low power can show both these two beautiful clusters in the same field of view. They are most easily found by sweeping with your eyes, binocular

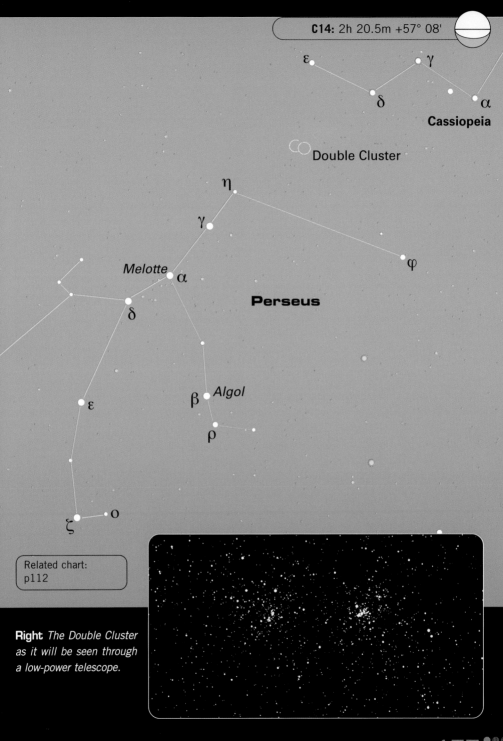

ε

γ

δ

α

Cassiopeia

Double Cluster

η

γ

φ

Melotte α

Perseus

δ

ε

β *Algol*

ρ

ζ ο

Related chart:
p112

Right *The Double Cluster as it will be seen through a low-power telescope.*

157

or finder scope to the east and a little south of Cassiopeia, following the line set by its bright stars Gamma (γ) and Delta (δ). The bright cores of the two clusters are separated by just less than one Moon diameter, 25 arc minutes, and together they cover over a degree of sky. Given their separation and individual visual brightness of between 4th and 5th magnitude, one should be able to see them as separate entities. But this is not usually the case. Surprisingly perhaps, the best chance to do so is by observing them just as twilight ends; when they first appear to the eye but the background stars of the Milky Way are still invisible. (In a similar fashion, the brighter stars of constellations – those that form the patterns that we learn – show up far more clearly under twilight or light-polluted conditions than when seen in really dark conditions. This is why you are advised to learn the shapes and locations of the constellations when the sky conditions are not too good!) The two clusters, also known as h and Chi (χ) Persei, are a beautiful sight in 10x50 binoculars; each cluster having a bright centre and many individually resolved stars. With a low-power eyepiece both can be seen in the same field and then, moving up to medium power, each can be observed in detail. They lie in the Perseus spiral arm of the Milky Way some 7300 light years away, and were both formed about 3 million years ago.

Perseus

Algol – Beta (β) Persei
• Eclipsing binary   E B

Algol is one of the most remarkable and most famous individual stars in the sky. Its Arabic name is Al Ghul, which means the 'demon' star (Ghul is related to 'ghoul', a ghost). Why a demon? Because it winks! Every 2.87 days its brightness quickly drops from magnitude 2.1 to 3.4 and then rises again to 2.1 over a period of 10 hours. John Goodricke of York was one of the first astronomers who discovered its regular brightness variations in 1782-3. Much later, in 1881, astronomers realized that the effect could be caused by a binary system in which the orbital plane of the two stars was almost in line with the Earth, so that every 2.87 days there is a partial eclipse! This is when the fainter star of the two comes in front of the brighter. In between each major drop in brightness, there is a much smaller drop as the brighter star comes in front of the fainter. The primary star is a blue B-type star with a surface temperature of 12,000K. The secondary is a much larger but dimmer K-type orange giant star. Interestingly, the two stars do not seem to be following the normal rules of stellar evolution. More massive stars evolve faster than less massive ones, so the orange giant – which has evolved away from the main sequence – should be more massive than the blue primary star. But it has less mass! It appears that material may be flowing from the giant star (so reducing its mass) onto the normal star whose mass is thus increasing.

Algol: 3h 8.2m +40° 57'

The constellation
Sagittarius

Lying in the direction of the centre of the Milky Way, Sagittarius is one of the richest constellations in the sky. Sadly, for observers in northern Europe, it tends to be lost low in the southern horizon during the short summer months. An excellent reason to travel to the southern hemisphere to observe it where it is high in the sky during the winter! Its main stars form the shape of a teapot. It contains 15 Messier objects, four of which have been included in the A-List.

M8   Sagittarius

**The Lagoon Nebula • Diffuse
nebula and open cluster   B L M**

This is perhaps one of the most beautiful regions of the milky way. It lies just over 5° to the west of and slightly north of Lambda (λ) Sagittarii (the top of the teapot's lid). So placing Lambda on the appropriate side of a binocular or finder field of view should bring M8 into the other side of the field. It is very hard to miss and, at about 5th magnitude, can be picked out using your eyes alone. The region is one and a half degrees wide with the main region of bright nebulosity at one end and a young open cluster, NGC 6530, towards the other. A dark east-west dust lane, most obvious at low and medium powers, may be why it was called the Lagoon Nebula, though it looks more like a river than a lagoon. The gas that glows in the bright emission regions is excited by the ultraviolet light from a very hot 6th

Above *The rich star fields of Sagittarius, which lie toward the centre of our galaxy, contain many open clusters of young stars.*

magnitude star, 9 Sagittarii, and others, most of which are hidden by dust. The very brightest part of this region is named the 'hourglass nebula' after its shape. Though not seen in small telescopes, the nebula contains many small dark compact 'globules'. These are collapsing clouds of dust and gas, typically 10,000 AU across, which are new stars in their first stages of formation.

M17   Sagittarius

The Swan or Omega Nebula
• Diffuse nebula   B M

M17 has many names as well as the two most common ones, given above. It is also called the Horseshoe or Lobster nebula – the latter particularly in the southern hemisphere. As in all bright nebulae, it is shining as a result of gas being excited by the ultraviolet light emitted from the hot young stars embedded within it. It is found right at the northern edge of Sagittarius on the boundary with Scutum, 9° north and a little west of the star Lambda (λ) Sagittarii, that delineates the top of the teapot's lid. So, start with this star at the left (Southern observers: right) of the field of a pair of binoculars or finder scope and move northwards from this star by about two field widths. This 6th magnitude patch of light, somewhat larger than the size of the Full Moon, is easily visible in binoculars and can even be picked out by the unaided eye under dark skies. A telescope shows the swan-shaped central region of the nebula while among fainter wisps of nebulosity, below the swan, can be seen an open cluster of 9th magnitude blue stars. The brightest part of the nebula is about 15 light years across and contains sufficient gas to form hundreds of stars.

M17: 18h 20.8m -16° 11'

M20 • M21   Sagittarius

Trifid Nebula • Diffuse nebula
and cluster   B L M

M20 lies just north of the Lagoon nebula (see above as to how to find it) and will be encompassed with both M8 and the open cluster M21 in a single 2° field – as can be observed by most small telescopes at low power. It was discovered by Charles Messier in 1764. He described it as a cluster of 8th to 9th magnitude stars within an envelope of nebulosity. It is called the Trifid Nebula, as it is seemingly split into three segments by dust lanes that radiate from the bright central region. It also looks like a clover leaf. Colour photographs show that as well as the emission nebulosity that appears red in photographs (emitted by excited hydrogen), there is also an extensive blue reflection nebula on its northern side. Moving north-east by 1° from M20 is M21, a 6.5 magnitude open cluster containing about 60 stars. There is a strong concentration towards its centre. The brightest stars are of type B0 – giant stars that have very short lives – so we know that this cluster must be very young, perhaps only 4.6 million years old.

M20: 18h 02.6m -23° 02'
M21: 18h 04.2m -22° 30'

Eagle Nebula

Swan Nebula

μ

Trifid Nebula

Lagoon Nebula

λ

σ

φ

τ

δ

ζ

γ

The Teapot

ε

Sagittarius

M7

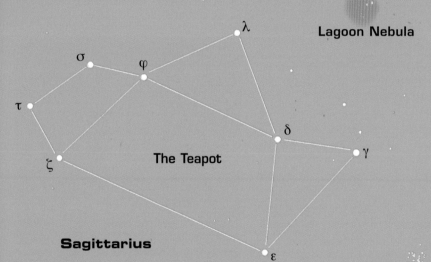

Related chart:
p111

The constellation
Scorpius

Scorpius, with its tail arching below Sagittarius, includes Antares, the 13th brightest star in the sky. An irregular variable, its brightness varies from 1.02 to 0.86 magnitudes. Note that the constellation is called Scorpius not Scorpio – as used by astrologers! It lies in the plane of the Milky Way galaxy and is thus rich in both globular and open star clusters. Four of the latter are included in the A-List. It is low in the south for Northern observers during summer, but high in the winter skies for those in the southern hemisphere.

M6   Scorpius

The Butterfly Cluster
• Open cluster   E B L

M6 is a cluster of about 120 stars that suggests the outline of a butterfly with open wings. The brightest stars in the cluster are of magnitudes 6 to 7, with 60 greater than 11th magnitude. The brighter stars are blue-white with the exception of the very brightest of all; a yellow-orange semi-regular variable star that alters in brightness from 7 up to 5.5 magnitudes every 850 days or so. It makes a very nice colour contrast with the blue-white stars around it. Overall, the visual magnitude of the cluster is 4.2 so it should be easily visible to the unaided eye under dark skies. It lies 4.5° north-west of M7 (*see* below) and is probably found most easily in binoculars or finder scope by scanning north-

Right *The dust lanes that run through the Milky Way are seen in this image of the rich star fields in Scorpius.*

west from M7 when both will be seen in the same field of view. M6 extends over an area some 20 arc minutes across; two-thirds the angular diameter of the Moon. Its distance is difficult to estimate due to the obscuration of light by the dust in the Milky Way but is approximately 1600 light years away, which would make the cluster extend across a volume of space 12 light years in diameter. It is estimated to be 100 million years old.

M7   Scorpius

Ptolemy's Cluster • Open cluster
E B L

This is by far the most obvious cluster in Scorpius, standing out in a very bright region of the milky way. Just imagine that the 'teapot' of Sagittarius, to the north-east, was pouring tea.

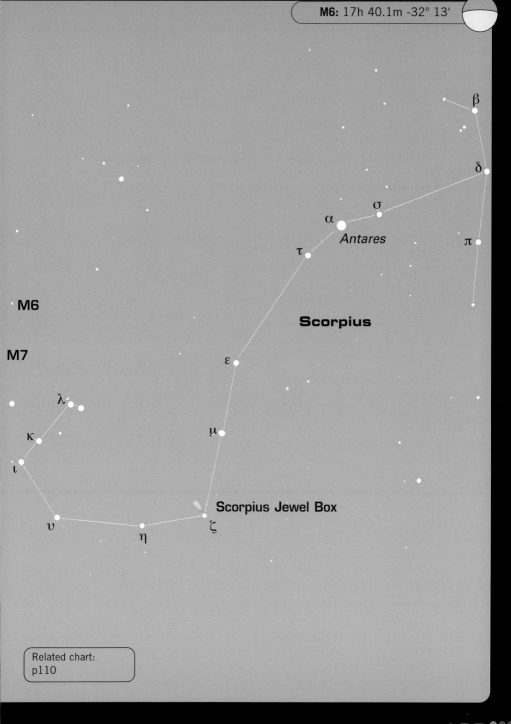

β

δ

σ

π

α
Antares

τ

M6

M7

Scorpius

ε

λ

κ

μ

ι

Scorpius Jewel Box

υ

η

ζ

Related chart:
p110

M7 is just where you would put the cup! It is obvious in binoculars and, with an overall visual magnitude of 3.3, easily visible to the unaided eye. A splendid cluster, it was mentioned by Ptolemy in 130AD and it has been suggested recently that it be named after him – an excellent idea. The cluster contains about 80 stars spread over an angular field 80 arc minutes across and all shine brighter than 10th magnitude. Its brightest member is a yellow giant of 5th magnitude while a dozen or so are brighter than 7th magnitude. It is thought to be over twice as old as its neighbour M6, which makes it about 220 million years old.

M7: 17h 53.9m -34° 49'

C76   Scorpius

The Scorpius or Northern Jewel Box • Twin open cluster   E B L

This is a name given to yet another wonderful region of the milky way at the western-end of the tail of the scorpion, almost due south of Epsilon (ε) Scorpii. Sweep south from Epsilon Scorpii by approximately two binocular fields of view, passing on the way a pair of 3rd magnitude stars Mu^1 (μ1) and Mu^2 (μ2) and stop as Zeta (ζ) Scorpii, another double star system, enters the field of view. Interestingly, Zeta is one of the most luminous stars in our galaxy – over 100,000 times as bright as our Sun! Just north of Zeta (ζ) Scorpii is a bright and compact open cluster NGC 6231, north-east of which are two loose open clusters Cr 316 and Tr 24 – numbers 316 and 24 respectively in the catalogues of clusters compiled by P. Collinder and R. J. Trumpler in the 1930s. Continuing north is the second compact cluster NGC 6242. A bright region of nebulosity called IC4268 lies just to the north of the loose clusters. This region has all the appearances of

a comet – the 'tail' arching north-west away from the 'coma' formed by NGC 6231 – so the region is also called the 'False Comet' Nebula. It marks one of the Milky Way's nearby spiral arms, some 6000 light years distant. NGC 6231 is a very young cluster, perhaps 3.2 million years old. This is indicated by the fact that its brightest star is an O-type star of apparent magnitude 4.7. Such stars burn up their fuel very quickly and so have very short (but dramatic) lives!

C76: 16h 54.2m -41° 50'

The constellation
Taurus

Taurus is highest in the sky during December and January, but easily visible in neighbouring months for those in the northern hemisphere. It forms part of a spectacular skyscape along with Orion, Canis Major and Gemini. Taurus contains three A-List objects; the two closest open clusters to us in the heavens and also a supernova remnant.

C41   Taurus

The Hyades Cluster • Open cluster   E B L

This famous V-shaped cluster outlines the head of Taurus, the Bull, with the bright star Aldebaran as his eye. In fact, Aldebaran is not part of the cluster and lies at only half of its 150 light-year distance. Stars can be shown to be part of a cluster by measuring their 'proper motion' – the path they take across the sky. All the stars of the Hyades are moving along paral-

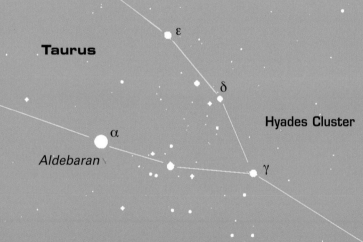

Taurus

ε

δ

Hyades Cluster

α

Aldebaran

γ

Related chart:
p107

lel paths eastwards towards a point close to Betelgeuse in Orion. In contrast, Aldebaran is moving southwards. The cluster has a central core about 10 light years in diameter with outer members spread over a volume some 80 light years across. It is thought to be about 730 million years old. Its angular extent is over 5° across so is best seen with binoculars, but a telescope at low power can give a nice view of its core including a prominent double star.

M45 — Taurus

The Pleiades Cluster
• Open cluster — **E B L**

Perhaps the most beautiful open cluster in the heavens, it is often called the 'Seven Sisters'. This name comes from a mythological story in which seven sisters were placed in the heavens to overcome their grief over the death of their father. Interestingly, almost no-one would ever see precisely seven stars. Most people actually see six stars with their unaided eye under dark skies. There is a significant drop in brightness between the sixth brightest star and the next four which still lie above the nominal unaided eye limit of 6th magnitude. So those that see more than six are likely to see nine or 10. Amazingly, under superb conditions with well dark-adapted eyes some have managed to discern up to 30! The cluster is about 2° across and binoculars or a short focal length telescope at low power can encompass the whole of the cluster – one of the best sights in the sky. Particularly pretty is a little triangle of stars just besides Alcyone, the brightest star in the cluster, and a double star appearing almost in the centre of the cluster. The Pleiades cluster is very young, only 70–100,000,000 years old, and

β

Crab Nebula

ζ

Related chart:
p107

contains about 100 stars. It is just 380 light years away from us, so is a relatively close neighbour in space. Colour photographs show that the brighter stars of the cluster are surrounded by blue reflection nebulae. These are caused by starlight reflecting off dust grains in the space between the stars and so, not surprisingly, is most apparent close to the three brightest stars. If the sky is very transparent and dark, this nebulosity can be observed, but almost in reverse: the background sky at the very centre of the 'square' formed by the four brightest stars of the cluster appears distinctly darker than that close to the stars themselves.

M45: 04h 26.9m +15° 52'

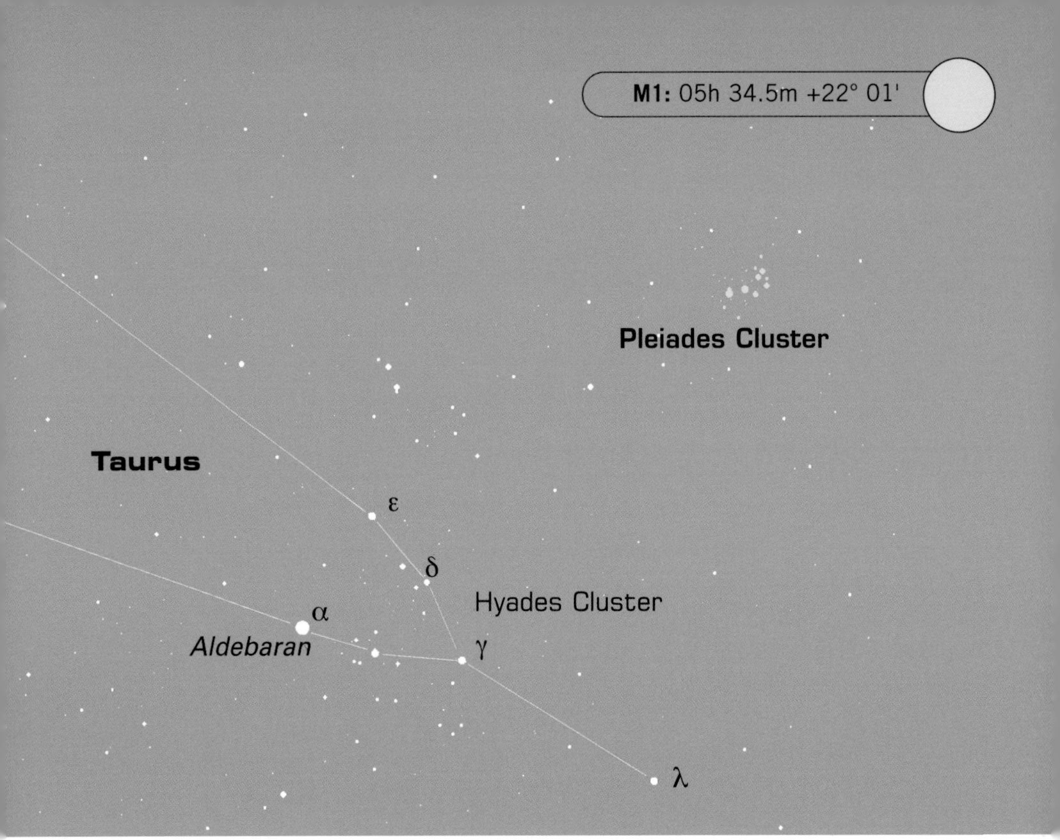

Pleiades Cluster

Taurus

ε

δ

Hyades Cluster

α

Aldebaran

γ

λ

M1 — Taurus

The Crab Nebula
• Supernova remnant — M

This is one of two supernova remnants in the A-List (the other being the Veil Nebula in Cygnus) and is one of around 100 such objects known of in our galaxy. The progenitor star was seen to explode in AD1054 and is recorded in Chinese texts. There are no European records perhaps because, in the eyes of the church, the heavens were meant to be perfect and unchanging so the scribes, who were generally monks, could well have been reluctant to chronicle such events. Since the explosion, the gas that was ejected as the star exploded has been expanding out into space and is gradually getting fainter. The fact that it still glows in the visual part of the spectrum at all is due to a powerhouse at its centre – the Crab Pulsar. The core of the massive progenitor star collapsed down until it was no bigger than the size of a big city – but still weighing more than our Sun – and began to spin rapidly. Energy is radiated away from the region of space above its magnetic poles and it acts like an interstellar light house; two beams of light and radio waves sweeping around the sky 30 times a second. Some of the energy radiated away is keeping the nebula glowing. The nebula is a 6 by 4 minute of arc patch of nebulosity just 1° northwest of the 3rd magnitude star Zeta (ζ) Tauri. Binoculars or a telescope at low power can thus show both in the same field so, given dark skies, M1 is easy to find with 8x40 or 10x50 binoculars. Higher magnification using a telescope may make it somewhat easier to see. The pulsar is one of two stars at the centre of the nebula, both about 16th magnitude. Given a 10in (250mm) telescope and a superb, dark sky, observing site with excellent seeing conditions it can even be observed – the most exotic object anyone could ever expect to see. Its name, the Crab Nebula, was given to it by the sixth Earl of Rosse who observed it with a 72-inch reflector, then the biggest telescope in the world, from Birr Castle in Ireland. In his drawing it appears somewhat like a horseshoe crab (some would say pineapple), hence the name.

The constellation
Triangulum

See p122' for Andromeda and Triangulum'

The constellation
Tucana

This small constellation, close to the south celestial pole, is best seen in springtime. Though small, with few bright stars, it contains two of the jewels of the southern sky

SMC   Tucana

The Small Magellanic Cloud
• Irregular galaxy   E B L

This is a small companion galaxy to our own Milky Way. It lies at a distance of 210 to 250 thousand light years, making it the third nearest galaxy to us (after the Large Magellanic Cloud and a dwarf elliptical galaxy in Sagittarius). It covers an angular extent of 280 by 160 arc minutes appearing as a faint cloud against a dark sky. The SMC must have been known since antiquity but it was 'discovered' by the explorer Ferdinand Magellan in 1519. To the unaided eye it appears like a little piece of the milky way. Binoculars and telescopes will show that it contains both open star clusters and bright nebulae. The SMC contains a large number of hot blue stars. As these have a relatively short life it shows that the SMC has recently undergone a period of star formation. The SMC follows a nearly circular orbit around our galaxy. Knowing the period and diameter of the orbit enables a calculation to be made of the total mass of the Milky Way galaxy. This turns out to be substantially more than the calculated mass of normal matter (in the form of stars, planets, gas and dust) and thus indicates

Left *The Small Magellanic Cloud is a small irregular galaxy orbiting the Milky Way which is seen in the sky near the south galactic pole.*

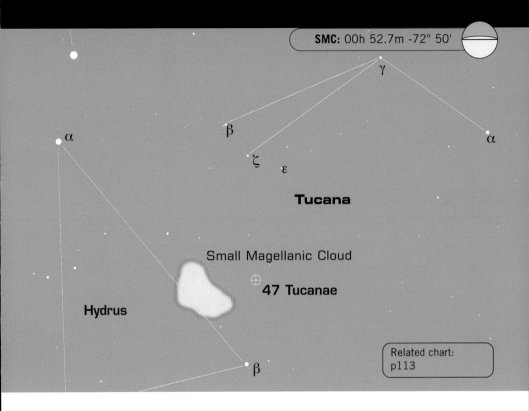

γ

β

ζ

ε

α

α

Tucana

Small Magellanic Cloud

⊕ **47 Tucanae**

Hydrus

β

Related chart:
p113

---

the presence of 'dark matter' – a form of matter about which little is yet known.

C106   Tucana

47 Tucanae • Globular cluster
E B M

Close to the SMC lies the second largest and brightest globular cluster in the sky. With a visual brightness of very close to 4, it is easily seen with the unaided eye under dark skies – its stars covering an area about the size of the Full Moon. It ranks second to Omega Centauri but it is only 0.05 magnitudes fainter – a virtually insignificant difference. Both 47 Tucanae and Omega Centauri exceed the brightness of any other globular clusters circling our galaxy by a factor of three. It was

not 'discovered' until 1751 when it was observed by Lacaille, but had been charted as a 'star' in Bayer's 1603 Uranometria. (It was Bayer who gave the brighter stars their Greek letter designations; *see* p20.) 47 Tuc (its common abbreviation) lies just over 13,000 light years from the Sun – one of the closest globular clusters to us. Its brightest stars are about 14th magnitude so they can be resolved with an 8in (200mm) telescope as individual points of light against the diffuse background glow from fainter stars in the cluster. It has a well defined compact core. Globular clusters are composed of some of the oldest stars in our galaxy – some estimates give them an age of over 12 billion years! This helps put constraints on our models of the evolution of the universe. It MUST be older than the oldest stars!

C106: 00h 24.1 -72° 05'

The constellations
Ursa Major and Canes Venatici

Sadly, these two constellations are only visible to northern latitude observers. The 'Plough' or 'Big Dipper' – shown by the thicker lines in the chart – is one of the best known groups of stars in the sky and almost everyone knows that by following the line of the two stars, Merak and Dubhe (known as the pointers), that make up the right hand side of the plough, brings you to the Pole Star, hence showing where true north is. The stars that make up the outline of the bear are not quite so obvious! Ursa Major contains two A-list objects along with some other interesting objects that are also shown on the chart. Canes Venatici is a very small and insignificant constellation that lies below the tail of the bear; but it does contain one of the most beautiful objects in the sky, the Whirlpool Nebula.

Alcor • Mizar — Ursa Major

Visual and telescopic doubles
E B M

Mizar, Zeta (ζ) Ursa Majoris, is the middle star of the three stars that make up the tail of the bear. With good sight, you should be able to see that it has a close neighbouring star – 12 arc minutes away – which is called Alcor. The pair, easily seen in binoculars, together are called the 'horse and rider'. The two stars orbit each other about once every million years and, in 1650, were the first double star ever to be observed. If however one now observes the pair with a telescope it will be seen that Mizar is itself a double star comprising two white stars 14 arc seconds apart – easily split even when seeing conditions are not good. In fact (though not visible in any telescope) each of the two stars is itself a double star so Alcor and Mizar make up a quintuple star system! In a telescope field using a medium power eyepiece with both Alcor and Mizar in view, a third faint reddish star will be seen to form a flattened triangle with them. Called Sidus Ludovicianum, it was named after Ernst-Ludwig V by a (not very good) observer who mistook it for a planet!

M81 • M82 — Ursa Major

Spiral and irregular galaxies
B L M

These form a pair of galaxies that are close together both visually and in their actual separation. With no bright stars close by they are not the easiest to find. It is probably best to use the line formed by Gamma (γ) and Alpha (α) Ursa Majoris, the bottom left and top right stars of the plough, as a pointer. If one moves binoculars or a finder scope up and to the right along the line of these two stars, the two galaxies will be found 10° – or two field diameters – from α Ursa Majoris. Both galaxies lie at a distance of about 12 million light years and their centres are only some 150,000 light years apart. A telescope with a low power eyepiece will show both galaxies in the same field – they are 37 arc minutes apart – then higher powers can be used to observe each individually.

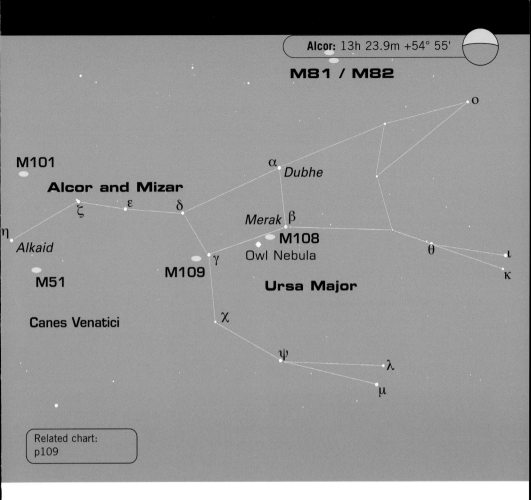

M81 / M82

M101

Alcor and Mizar

ζ   ε   δ

η

Alkaid

M51

M109

Canes Venatici

α
Dubhe

Merak β

M108
Owl Nebula

γ

χ

Ursa Major

ψ

μ

λ

ο

θ

ι

κ

Related chart:
p109

M81, at magnitude 6.8, is one of the brightest galaxies visible in the sky and thus easy to observe with a small telescope. It is a type Sb spiral galaxy, the same type as the Andromeda Nebula, M31, with a bright nucleus and reasonably open spiral arms. A small telescope will show the bright core of the nucleus and, if the sky is dark and transparent, a hint of the spiral arm structure particularly at the upper right and lower left.

M82 is an elongated irregular galaxy that is less bright overall, at magnitude 8.4. However, as it is smaller in area, it is still easy to make out as a thin cigar-like band of light. It seems that a 'recent' close encounter with M81 has initiated a major burst of star formation and it is thus known as a 'starburst' galaxy. Radio images of its core have shown many very young supernova remnants resulting from the explosive end to the massive stars that have been formed there.

M81: 09h 55.6m +69° 04'
M82: 09h 55.8m +69° 41'

Spiral galaxy in Canes Venatici

M

This is perhaps the finest example of a face on spiral galaxy that we can observe, certainly with an amateur telescope. It lies 3.5° south-west from the left hand star of the Bear's tail, Eta (η) Ursa Majoris, and forms a right angle triangle with it and Mizar. The distance of M51, a type Sc galaxy, is still in some dispute; some put it as near as 15 million light years, but a recent determination gave 31 million light years. It was originally discovered by Messier himself in 1773. Later Lord Rosse observed it with his giant 72in telescope at Birr Castle in Ireland and made a wonderful drawing showing the extended spiral arm structure – the first time spiral structure had ever been observed in a galaxy. In fact, M51 is a pair of interacting galaxies, NGC5194, the large

spiral, and NGC5195, its smaller and irregular companion. It is thought that the gravitational interaction of the smaller galaxy may have triggered the formation of the spiral arms in the larger. Under reasonably dark skies, a small telescope will show the nuclei of the two galaxies separated by 4.5 minutes of arc. Under really dark and transparent skies the spiral structure may just be discerned, particularly with telescopes of 8in (200 mm) or more in aperture.

Above *M51 with its companion, NGC5194, are interacting galaxies. M51 was the first galaxy in which spiral arms were observed.*

Other Messier objects

The chart also shows four other Messier objects; three galaxies and one planetary nebula. M108 is a 10th magnitude type Sc galaxy very close to Merak, Beta (β) Ursa Majoris, whilst M109 is a type SBc spiral, magnitude 9.8, close to Gamma (γ) Ursa Majoris. It has a prominent central bar of stars from the end of which flow the spiral arms – hence the 'B', for barred, in its Hubble classification SBc. M101 makes a nice triangle with Eta (η) and Zeta (ζ) Ursa Majoris at the end of the tail of the bear. It is a magnitude 7.9, type Sc, galaxy at a distance of 27 million light years. It is called the 'pin-wheel galaxy' due to its open arms, but a small telescope will probably only show the compact nucleus. Finally we have M97, the Owl Nebula. This is a planetary nebula close to M108. At magnitude 9.9 it will require a dark sky to be able to pick it out. The disk of glowing material has two prominent circular 'holes' which appear like the eyes of an owl so giving it its name.

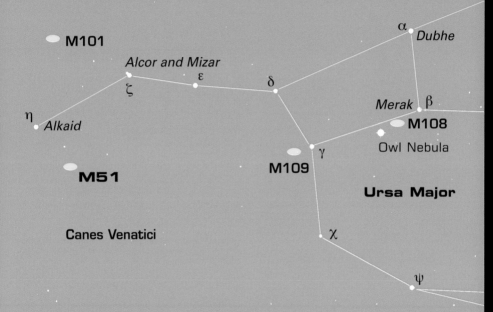

M101

Alcor and Mizar

ζ

ε

δ

α Dubhe

η

Alkaid

M51

Merak β

M108

Owl Nebula

γ

M109

Ursa Major

Canes Venatici

χ

ψ

Related chart:
p109

Virgo is a large constellation whose brightest star is 1st magnitude Spica but which contains few other bright stars. At the north of the constellation and crossing into Coma Berenices is a region of the sky called the 'realm of the galaxies'. It is in this direction that we are looking towards the centre of our local super-cluster of galaxies, the Virgo super-cluster of which our small 'local group' of galaxies is an outlying member. Here Messier logged 16 galaxies, and sweeping this region under dark skies is very rewarding. One galaxy from this region is in the A-List along with a second galaxy on the boundary with Corvus. A nice, but currently very challenging, double star system is also in the list. Virgo is highest in the sky from April to June.

M87   Virgo

M87 – Giant elliptical galaxy   M

M87 is a magnitude 8.6 giant elliptical galaxy – one of the largest galaxies in the universe – lying at a distance of 60 million light years at the centre of the Virgo cluster. There are no bright stars nearby and thus it is not too easy to find. This is compounded by the fact that it is surrounded by other galaxies, though these are a little less bright. It lies on the line between Beta (β)

Right *A Hubble Space Telescope image of the core and jet of M87. The star-like objects in the field are in fact globular clusters like M13 in our own galaxy.*

Leonis to the west and Epsilon (ε) Virginis to the east and is almost halfway between them just over 10°, about two finder fields of view, from β Leonis. If your telescope is equatorially mounted and equipped with an eyepiece whose actual field of view is 1.5° or more, then you can place (ε) Virginis at the top of the field (with inverted image) and lock the declination axis. (Southern observers: bottom of the field.) Then sweep westwards in right ascension by 31 minutes (7.75°) and M87 should lie towards the bottom (southern: top) of the field. M87 will appear as a roughly spherical fuzzy disc.

M87 is an exceptional galaxy. Though of similar diameter to our own Milky Way galaxy, being spherical rather than just a thin disc in shape, it contains vastly more stars, several trillion in number. Deep exposure photographs show that it extends far beyond the 7 arc minutes shown in most photographs, perhaps to half a degree across. It is surrounded by a swarm of globular clusters. M87 also has a jet visible on short exposure photographs. This jet is an outflow of material ejected from the region of a supermassive black hole at the centre of the galaxy. The

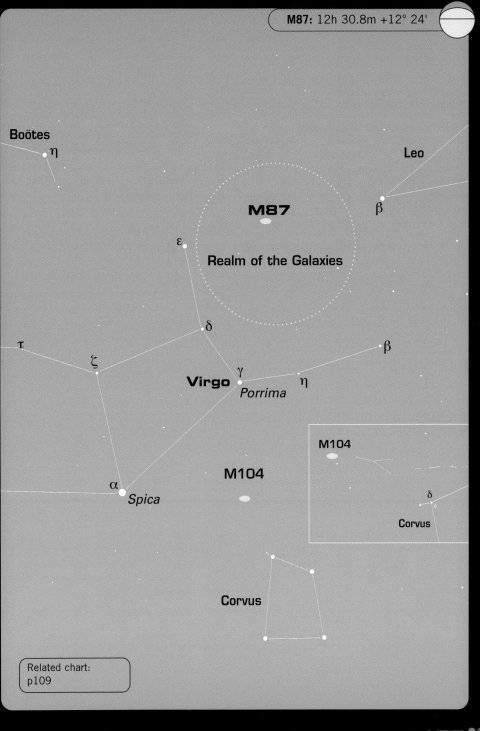

Boötes

η

Leo

M87

β

ε

Realm of the Galaxies

δ

τ

ζ

Virgo

γ

Porrima

η

β

M104

M104

α

Spica

δ

Corvus

Corvus

Related chart:
p109

jet is the source of intense radio emission and thus M87 is termed a radio galaxy and has been given the name Virgo A.

M104 — Virgo

Sombrero Galaxy • Spiral galaxy
M H

M104 is an 8th magnitude type Sa spiral galaxy seen almost edge-on. Type Sa spirals have a prominent nucleus surrounded by tightly wound spiral arms. It lies on almost the same declination as Spica, so centre Spica in the telescope field using a low power eyepiece, lock the declination axis and move westwards in right ascension by 45 minutes (RA is measured in time: 45 minutes corresponds to 11.25°). One can also reach it by moving north and a little east by about one finder's field of view from Delta (δ) Corvi. In addition, a line of 7th magnitude stars leads up towards M104 from Gamma (γ) Corvi reaching an arrow shaped asterism pointing a little to the west of M104. (See inset on star chart.) As with all galaxies, you really need a dark and transparent sky to observe it well. At low power it appears as a small oval patch of light. With higher powers the brilliant core becomes more obvious and a dark band, due to a prominent dust lane, appears to cross it – this dust lane forming the rim of the Sombrero Hat. Its dimensions are 9 by 4 arc minutes and M104 is thought to lie at a distance of 50 to 60 million light years.

Virgo

Gamma Virginis Porrima
• Double star — E H

Porrima is a magnitude 2.7 star lying 10° from Spica in the direct line from Spica towards Beta (β) Leonis – the star at the hindquarters of the lion. The view with a telescope under high power can be stunning: a pair of identical twins, each being type F stars, white in colour, whose surface temperature is about 7000K, somewhat hotter than our Sun. Lying at a distance of 38 light years, they are 50% more massive than our Sun and about 4 times greater in luminosity. The two stars are orbiting each other with a period of 170 years having a separation of 40 AU, about the distance of Pluto from the Sun. Their angular separation in the sky reaches a minimum in 2005 when they will be less than 1 arc second apart making them very difficult to split. You will need a night of excellent seeing and first-class telescope optics to succeed in the next few years! However it will be very worthwhile observing them each coming year as you will gradually observe them moving apart – out to a separation of about 2 arc seconds by 2012. There are very few double stars where one can see significant differences in just a few years! Their maximum separation, of just under six arc seconds, will occur around 2080.

Gamma: 12h 41.7m -1° 27'

OTHER MESSIER GALAXIES.
The region in the circle on the sky with Epsilon (ϵ) Virginis and Beta (β) Leonis at either side (shown on the chart) will reveal a host of galaxies when observing with a well dark-adapted eye under dark skies. Scanning a telescope at medium power should enable you to see many other galaxies in Messier's catalogue. From north to south these are: M85, M100, M98, M91, M88, M99, M90, M86, M84, M89, M87, M58, M59, M60, M49 and M61.

M104: 12h 40.0' -11° 37'

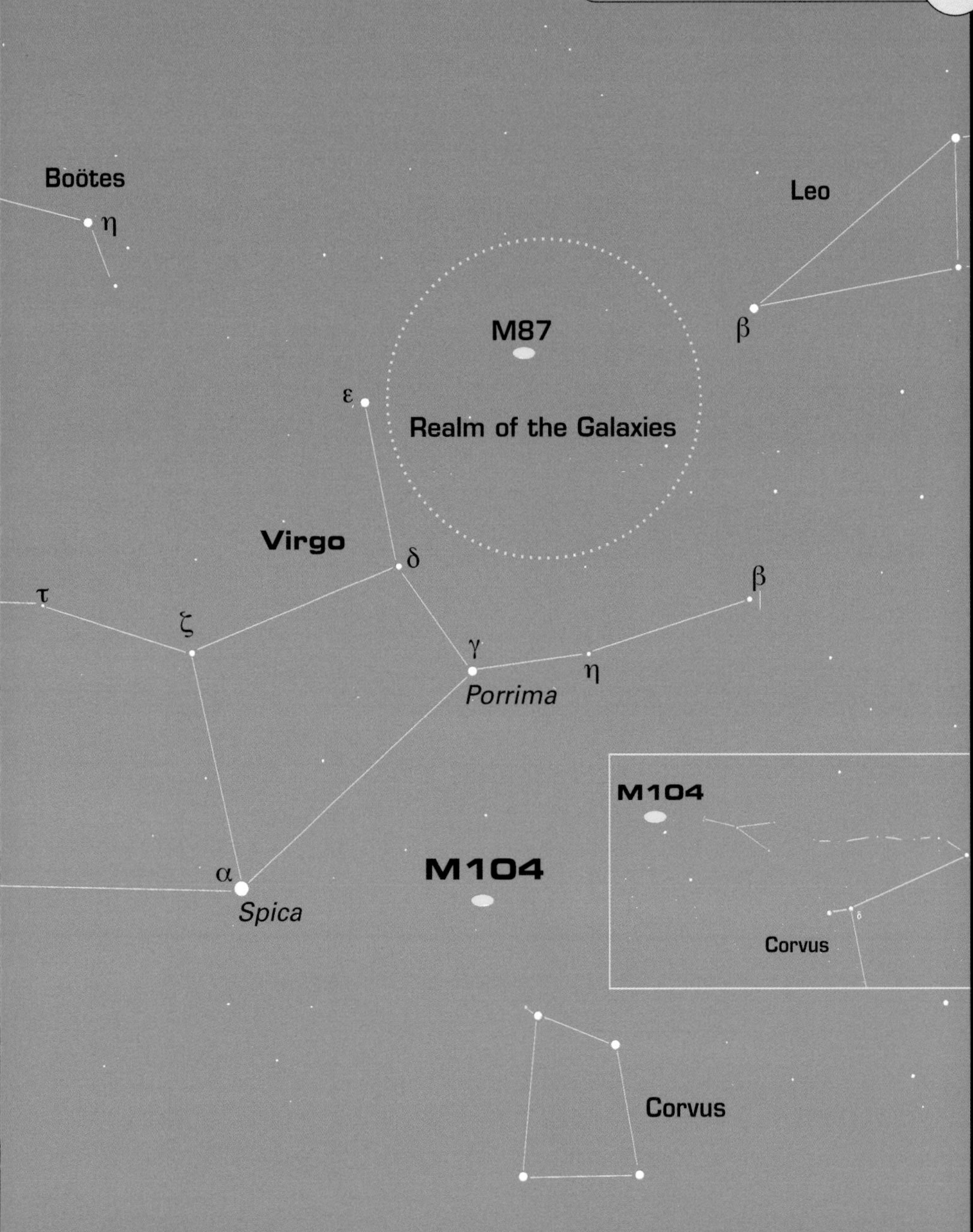
Boötes
η
Leo
β
M87
Realm of the Galaxies
ε
Virgo
δ
β
τ
ζ
γ
η
Porrima
M104
Corvus
α
Spica
M104
Corvus

Related charts:
p109

The constellation
Vulpecula

Vulpecula is a small constellation that lies between Cygnus and Lyra to the north and Delphinus, Sagitta and Aquila to the south. It has no bright stars but does, however, contain two A-List objects! It comes highest in the sky during July, August and September.

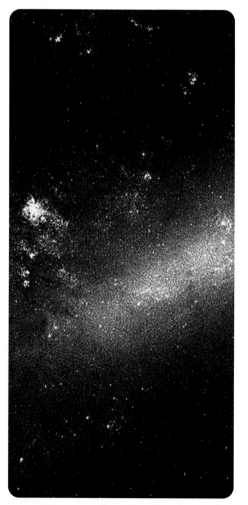

Brocchi — Vulpecula

Brocchi's Cluster • Open cluster or asterism — B L

Brocchi's Cluster is commonly called the 'Coathanger' because that is exactly what it looks like! Seen upside-down with binoculars from the northern hemisphere — the right way up for Southern observers — it is best found by sweeping across the milky way from the bright star Altair, in Aquila, one third of the way towards Vega, in Lyra. It should be easily spotted against a darker region of the Milky Way called the Cygnus Rift. It is a grouping of around 40 stars, but only 6 or so of the brighter ones have a common motion through space — the definition of the stars forming a cluster — so it might better be termed an 'asterism' - a chance pattern of stars. It subtends 1° across, so binoculars or a telescope at low power will show it best. There is a very nice colour contrast between two red giant stars forming part of the handle and the whiter stars of the 'bar'.

M27 — Vulpecula

Dumbbell Nebula • Planetary nebula — B M

In 1762 Charles Messier discovered the 27th object in his catalogue, describing it as an oval nebula without stars. Later John Herschel gave it its common name, the Dumbbell Nebula. It was the first planetary nebula to be discovered and is the remnant of a giant star that exploded at the end of its life leaving a cloud of dust and gas

Left *A section of the Large Magellanic Cloud showing the Tarantula Nebula in which a supernova was observed in 1987. (See p142)*

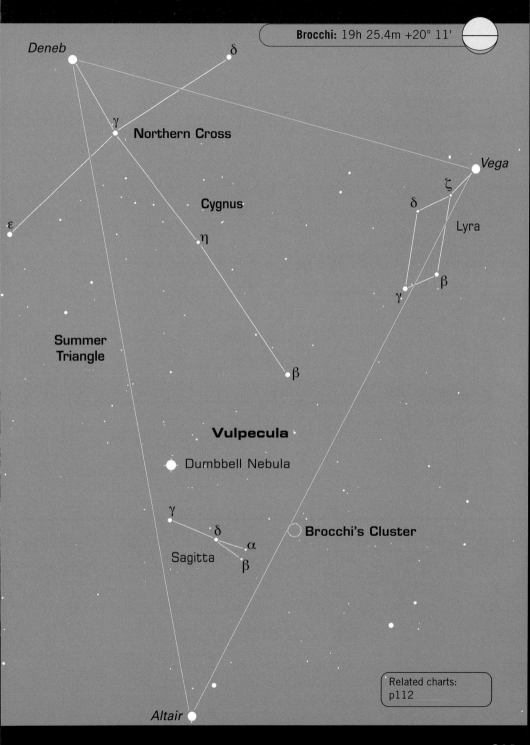

Deneb

δ

γ

Northern Cross

Vega

ζ

δ

Cygnus

Lyra

ε

η

β

γ

Summer
Triangle

β

Vulpecula

Dumbbell Nebula

γ

δ

Brocchi's Cluster

α

Sagitta

β

Altair

Related charts:
p112

surrounding the dying ember of its core – a so called white dwarf star about the size of the Earth. This white dwarf is still extremely hot, about 85,000K, and so is emitting ultraviolet light which is exciting the surrounding gas shell and making it glow. As with all nebulous objects, it is best searched for in dark and transparent skies. The easiest way to locate it is to first find Gamma (γ) Sagitta, the 3.5 magnitude star at the head of the tiny flattened triangle making up the 'arrow' of Sagitta. This lies 10° directly north of Altair in Aquila. If γ Sagitta is now placed at the bottom of a binocular or finder field of view, M27, which is 3° north, should be visible towards the top. (Southern observers put γ Sagitta at the top and look towards the bottom.) The nebula is 8 by 6 arc minutes in angular size and 7.4 in magnitude so it should be picked up as a bright spot in 8x40 or 10x50 binoculars. A telescope at medium power will show it as an elongated nebula, rather like an apple core in shape, and an 8in (200mm) telescope or larger may well show the 13.5 magnitude central white dwarf star. Its distance is not well known, perhaps 1,250 light years, and measurements of its rate of expansion, some 6.8 arc seconds per century, gives an estimated age of three to four thousand years.

Below *The Dumbbell Nebula. The central white dwarf star is surrounded by a shell of excited gas forming a planetary nebula.*

Vulpecula

Dumbbell Nebula

γ

δ

α

Sagitta

β

Brocchi's Cluster

β

Altair

Related charts:
p112

7

GLOSSARY
AND INDEX

Left *An artists interpretation of our solar system showing the angled orbit of Pluto, with the main inner orbits zoomed out into the foreground at left, with Mercury closest to the Sun, followed by Venus, the Earth-Moon system and Mars.*

Arc minute 1/60th of a degree of arc.

Arc second 1/60th of an arc minute.

Asteroid belt The band of rocky objects orbiting the Sun between Mars and Jupiter.

Astronomical unit (AU) The average distance between the Sun and the Earth, equal to 149,597,870,691km or about 150 million km.

Black body A hypothetical 'perfect' object which absorbs and re-radiates all the energy that falls on it. The radiation from a star is approximately that of a black body.

Celestial sphere The sky all around us gives the impression that we are at the centre of a massive hollow celestial sphere. An imaginary grid drawn on this sphere gives us celestial co-ordinates such as Right Ascension (longitude) and Declination (latitude).

Dark matter A study of the motions of stars within galaxies shows that, from the observed gravitational effects, the galaxies must contain some 10 times more matter than can be seen. The nature of this 'missing mass' is not known and is called dark matter.

Doppler effect The motion of an object toward or away from the observer results in compression or stretching of the waves of radiation from that object. One effect is that lines in the spectrum are shifted toward the red (red shift) if the object is moving away from the observer or toward the blue (blue shift) if the object is approaching the observer.

Dwarf (white, red) A small star. Red dwarfs are stars of mass less than about 0.5 x solar mass. They lie on the main sequence of the Hertzsprung-Russell diagram, slowly burning hydrogen. The universe is not old enough for any red dwarfs to have evolved beyond this. A white dwarf is a star at the end of an exciting life. It has a mass about the same as the Sun but its size is more like that of a planet.

Ecliptic The Sun's path across the sky. The Moon and planets are all found close to the ecliptic. The band of constellations that straddle the ecliptic is called the zodiac.

Equinox literally meaning 'equal nights'. Twice a year, in March and September, when the Sun crosses the celestial equator, the lengths of day and night are equal everywhere on Earth.

Hertzsprung-Russell diagram A plot of the absolute brightness or luminosity of a star against its colour or temperature. The diagram is extremely useful for describing how stars of different mass evolve over their lifetime.

Inferior/superior planet A planet whose orbit of the Sun is within or outside that of the Earth respectively.

Interstellar medium The material that lies between the stars. It is a mixture of mostly hydrogen gas and 'dust', the debris from previous generations of stars.

Kelvin scale of temperature The scale whose zero point is 'absolute zero' (-273.15°C) at which the motion of all particles ceases. The size of 1K = 1°C = 1.8°F so that, for example, the freezing point of water at 0°C = 273.15K = -32°F.

Luminosity The absolute brightness of an object.

Magnitude The brightness of a celestial object. The apparent magnitude is as it appears from Earth and absolute magnitude is if it were at a standard distance. A visual magnitude is the brightness in the part of the spectrum to which our eyes are most sensitive.

MAIN SEQUENCE The diagonal band in the Hertzsprung-Russell diagram which represents the longest time period in a star's evolution when stars of different mass are 'burning' their central hydrogen.

MERIDIAN The imaginary line passing between the north and south points on the horizon and the zenith.

NEWTON'S THEORY The simple theory in which there is a mutual attraction (or gravitational force) between any two bodies, the strength of which depends on the masses and the distance between them.

NOVA A 'new star' appearing in the sky where no bright star was known previously.

OCCULTATION An occultation occurs when one celestial body passes in front of another, thereby blocking out some or all of its light.

OPPOSITION A planet at opposition when it lies directly opposite the Sun in the sky.

PLANETESIMAL One of the building blocks of a planet.

POPULATION I, II STARS The grouping of stars in the galaxy by age. Population I are young 'metal-rich' stars found in the spiral arms of the galaxy; Population II are old 'metal-poor' objects found in the halo of the galaxy.

PRECESSION (OF EQUINOXES) The axis of the Earth spins like a top, moving around the sky with a period of 26,000 years. At present the North Pole almost coincides with the star Polaris. In about 12,000 years time it will be close to the star Vega in the constellation Lyra.

RADIAL VELOCITY A measure of the speed of an object either toward or away from the observer. See Doppler effect.

RETROGRADE Most planets and their satellites rotate in an anti-clockwise direction as seen from above the North Pole. Rotation in the opposite, clockwise, sense is said to be retrograde. Venus, Uranus and Pluto rotate in the retrograde sense.

SIDEREAL DAY The length of the day measured relative to the stars, equal to 23hr 56.1min.

SOLAR MINIMUM/MAXIMUM Times within the 11-year cycle of activity on the surface of the Sun.

SPECTRAL CLASS/SPECTRAL TYPE The visual classification of the appearance of the spectrum of a star.

STANDARD CANDLE A term used for an object whose absolute brightness is known and so can be used as a measure of the distance of its neighbours in space. Variable stars known as Cepheids are a good example of this.

SUPERNOVA The sudden brightening of a star by 10 magnitudes or more as a result of a massive explosion (*see* nova).

T TAURI PHASE An early phase in the life of some stars when they become very variable (*see* variable star).

VARIABLE STAR Most stars pass through 'variable' phases in their evolution when the light output varies, due to pulsation or perhaps an eruption. A star's brightness may also appear to vary if it is in orbit around another star.

ZENITH The point on the celestial sphere directly overhead.

Index

Constellation			
Tucana	C106	47 Tucanae	NGC 104
Andromeda	M31	The Andromeda Galaxy	NGC 224
Tucana	SMC	Small Magellanic Cloud	
Triangulum	M33	Triangulum Galaxy	NGC 598
Perseus	C14	η and Chi (χ) Persei	the double cluster
Perseus	Algol	Beta (β) Persei	HD 19356
Taurus	M45	Pleiades	Seven Sisters
Taurus	C41	Hyades	
Dorado	LMC	Large Magellanic Cloud	
Taurus	M1	Crab nebula	NGC 1952
Orion	M42	Orion Nebula	NGC 1976
Dorado	C103	30 Doradus	Tarantula Nebula
Auriga	M37	NGC 2099	
Gemini	M35	NGC 2168	
Canis Major	M41	NGC 2287	
Gemini	Castor	Alpha (α) Geminorum	HD 60178
Gemini	C39	Eskimo / Clown Nebula	NGC 2392
Cancer	M44	Beehive Cluster	Praesepe
Ursa Major	M81	NGC 3031	
Ursa Major	M82	NGC 3034	
Carina	C92	Eta (η) Carinae	NGC 3372 (nebula)
Leo	M95	NGC 3351	
Leo	M96	NGC 3368	
Leo	M65	NGC 3623	
Leo	M66	NGC 3627	
Crux	Acrux	Alpha (α) Crucis	HD 108248
Virgo	M87	NGC 4486	
Virgo	M104	Sombrero Hat	NGC 4594
Crux	C99	Coal Sack	
Crux	C94	(Herschel's) Jewel Box	NGC 4755
Virgo	Porrima	Gamma (γ) Virginis	HD 110379
Ursa Major	Mizar + Alcor	Zeta (ζ) + 80 Ursa Majoris	HD 116656/842
Centaurus	C77	Centaurus A	NGC 5128
Centaurus	C80	Omega (ω) Centauri	NGC 5139
Canes Venatici	M51	Whirlpool galaxy	NGC 5194/5
Centaurus	Alpha (α) Cen	Rigil Kentaurus	HD 128620
Hercules	M13	NGC 6205	
Scorpius	C76	Scorpius Jewel Box	NGC 6231
Hercules	M92	NGC 6341	
Scorpius	M6	Butterfly cluster	NGC 6405
Scorpius	M7	Ptolemy's cluster	NGC 6475
Sagittarius	M20	Trifid nebula	NGC 6514
Sagittarius	M8	Lagoon Nebula	NGC 6523
Sagittarius	M17	Swan Nebula	Omega Nebula
Lyra	Eps (ε) Lyrae	The double double	HD 173582/607
Lyra	M57	Ring Nebula	NGC 6720
Vulpecula	The Coathanger	Brocchi's Cluster	Collinder 399
Cygnus	Albireo	Beta (β) Cygni	HD 183912
Aquila	Eta (η) Aquilae	HD 187929	
Vulpecula	M27	Dumbbell Nebula	NGC 6853
Cygnus	C33/34	Veil Nebula / Cygnus Loop	NGC 6960/92/95
Pegasus	M15	NGC 7078	
Pegasus	51 Pegasi	HD 217014	

RA (2000)	Dec		Magnitude	Distance
00 24.1	-72 05	globular cluster	4	13,000
00 42.7	+41 16	spiral galaxy	3.4	2,900,000
00 52.7	-72 50	irregular galaxy		210–250,000
01 33.9	+30 39	spiral galaxy	5.7	3,000,000
02 20.5	+57 08	twin open clusters	4 + 5	7300
03 08.2	+40 57	eclipsing binary star	2.1 to 3.4	93
03 47.0	+24 07	open star cluster	2.9 (Alcyone)	400
04 26.9	+15 52	open star cluster	0.9 (Aldebaran)	150
05 23.6	-69 45	irregular galaxy		170–180,000
05 34.5	+22 01	supernova remnant	8	6500
05 35.4	-05 27	bright nebula	4	1600
05 38.6	-69 05	bright nebula + open	4	165–170,000
05 52.4	+32 33	cluster	6.2	4500
06 08.9	+24 20	open cluster	5	2700
06 46.0	-20 44	open cluster	4.5	2300
07 34.6	+31 53	open cluster	1.6	52
07 29.2	+20 55	multiple star system	~10	4000
08 40.1	+19 59	planetary nebula	3	577
09 55.6	+69 04	open cluster	6.8	12,000,000
09 55.8	+69 41	spiral galaxy	8.4	12,000,000
10 43.6	-59 52	irregular galaxy	6.2	>8000
10 44.0	+11 42	bright nebula +	9.7	38,000,000
10 46.8	+11 49	unstable star	9.2	41,000,000
11 18.9	+13 05	barred spiral galaxy	9.3	35,000,000
11 20.1	+12 59	spiral galaxy	8.9	41,000,000
12 26.6	-63 06	spiral galaxy	0.8	320
12 30.8	+12 24	spiral galaxy	8.6	60,000,000
12 40.0	-11 37	double star	8	50–60,000,000
12 52	-63 18	giant elliptical galaxy		2000
12 53.6	-60 21	spiral galaxy	4.2	7500
12 41.7	-01 27	dark nebula	2.7	38
13 23.9	+54 55	open cluster	2.3 + 4.0	80
13 25.5	-43 01	double star	7	10–16,000,000
13 26.8	-47 29	multiple star system	3.7	16,000
13 29.9	+47 12	active galaxy	8.4	15–30,000,000
14 39.6	-60 50	globular cluster	-0.3	4.4
16 41.7	+36 28	spiral galaxy	5.8	25,000
16 54.2	-41 50	multiple star system	2.6	6000
17 17.1	+43 08	globular cluster	6.5	27,000
17 40.1	-32 13	open cluster	4.2	1600
17 53.9	-34 49	globular cluster	3.3	800
18 02.6	-23 02	open cluster	8 to 9	5000
18 03.6	-24 23	open cluster	5	5200
18 20.8	-16 11	bright nebula	6	5000
18 44.3	+39 40	bright nebula	4.6 + 5.0	160
18 53.6	+33 02	bright nebula	8.8	2000
19 25.4	+20 11	multiple star	3.6	420
19 30.7	+27 57	planetary nebula	3.0 + 5.0	380
19 52.5	+01 00	asterism/open cluster	3.7 to 4.5	1300
19 59.6	+22 43	double star	7.3	1250
20 56.0	+31 43	cepheid variable star	~5	1400–2600
21 30.0	+12 10	planetary nebula	6.2	37,000
22 57.5	+20 46	supernova remnant	5.5	50
		globular cluster		
		planetary system		

Credits

AER Astronomy Educational Review (aer.noao.edu)

BAL The Bridgeman Art Library (A = Alinari; PC = Private Collection; L = Lauros; PM = Philip Mould Historical Portraits Ltd. London; OG = Orlicka Galerie, Rychnov na Kneznou; S =The Stapleton Collection)

CI Celestron International

FE Dr Fred Espenak

GG Gallo Images/gettyimages.com

G/M GreatStock/Masterfile

GPL Galaxy Picture Library (RS = Robin Scagell)

H NASA/Hubblesite

ME Mary Evans Picture Library

NA Nick Aldridge

NASA National Aeronautics and Space Administration (G = Galaxy)

O Orion Telescopes

RMPL Redferns Music Picture Library

SPL Science Photo Library (DA = David P. Anderson; JB = Julian Baum; CB = Chris Butler; DAH = David A. Hardy; NOAO = National Optical Astronomy Observatories; JS = John Sanford; USGS = US Geological Survey)

back cvr		NASA	46		CI	83	tr	NASA/G
endpapers	b	FE	47		FE	84	tr	BAL/S
1		NASA/G	48		ME	96	bl	AER
2–3		NA	51	tl	O	100	bl	AER
4–5		SPL/CB	51	tr	O	103	br	AER
6		NASA	52		O	114–115		FE
10	tr	GPL/RS	53		GPL/RS	116		SPL/NOAO
11	t	BAL/A	54	t	FE	118		NASA
	b	BAL/L	54	b	FE	119	b	AER
13	tl	BAL/PM	55	t	O	124	b	NASA
	bl	BAL/PC	55	m	O	128	b	NASA
14	tl	BAL/OG	55	b	O	130		NASA
15		FE	59		O	134		NASA
			64		NASA	138		H
16–17		G/M	64	b	SPL/JS	142		AER
18		FE	67		NASA	147		NASA
20		SPL/JB	70		SPL/DS SMU/NASA	149		NASA
27	tr	SPL/NASA	71	t	NASA	150		H
32		SPL/DH	71	b	SPL/NASA	152		H
33		H	73	b	GPL	157		NASA
34		H	75	t	SPL/USGS	159		NASA
38		H	75	m	GPL/JPL	162	bl	AER
39		NASA	75	b	GPL/JPL	162	br	NASA
40		H	76	br	SPL/NASA	168		NASA
41		H	79	tr	SPL/NASA	172		NASA
42		H	79	bl	GPL/JPL	174		NASA
43	t	FE	79	br	NASA	178		NA
43	b	H	81	tr	H	180		NASA
44		GG	82		GPL/JPL			